AF598078

APPLICATIONS OF DESIGN FOR MANUFACTURING AND ASSEMBLY

Edited by **Ancuța Păcurar**

Applications of Design for Manufacturing and Assembly
http://dx.doi.org/10.5772/intechopen.75475
Edited by Ancuța Păcurar

Contributors

Prashant Gangidi, Roy Lopez-Sesenes, Martha Roselia Contreras-Valenzuela, Alejandro David Guzmán-Clemente, Diana Yesenia Lagunas, Jose Gerardo Vera Dimas, Viridiana Aydee Leòn Hernandez, Alina Martinez Oropeza, Alber Eduardo Duque Alvarez, Khumbulani Mpofu, Ilesanmi Afolabi Daniyan, Ruta Miniotaite, Jonathan Becedas, Andrés Caparrós, Ancuta Carmen Păcurar

First published in London, United Kingdom, 2019 by IntechOpen
IntechOpen is the global imprint of INTECHOPEN LIMITED, registered in England and Wales, registration number: 11086078, The Shard, 25th floor, 32 London Bridge Street
London, SE19SG – United Kingdom
Printed in Croatia

British Library Cataloguing-in-Publication Data
A catalogue record for this book is available from the British Library

Additional hard copies can be obtained from orders@intechopen.com

Applications of Design for Manufacturing and Assembly, Edited by Ancuța Păcurar
p. cm.
Print ISBN 978-1-78984-935-6
Online ISBN 978-1-78984-936-3

For EU product safety concerns:
IN TECH d.o.o., Prolaz Marije Krucifikse Kozulić 3, 51000 Rijeka, Croatia,
info@intechopen.com or visit our website at intechopen.com.

Meet the editor

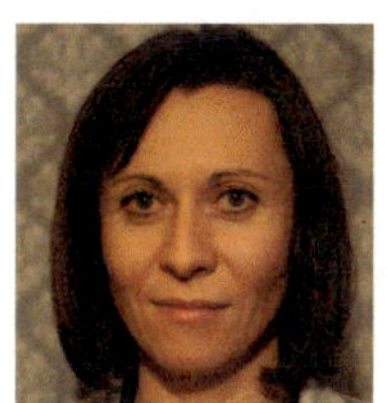

Dr. Eng. Ancuţa Păcurar is a lecturer at the Department of Manufacturing Engineering from the Faculty of Machine Building at the Technical University of Cluj-Napoca and is a member of the National Research Center for Innovative Rapid Manufacturing. She holds a PhD title in the Industrial Engineering field, with more than 50 articles presented at different international scientific conferences or indexed in significant international databases or ISI, and is involved as a member within several national and international projects. Her areas of expertise are manufacturing engineering, assembly technologies, maintenance system, additive manufacturing and design for environment. She is a member of the University Association of Manufacturing Engineering, Romania.

Contents

Preface IX

Chapter 1 **Introductory Chapter: Applications of Design for Manufacturing and Assembly 1**
Ancuța Carmen Păcurar

Chapter 2 **Application of Design for Manufacturing and Assembly: Development of a Multifeedstock Biodiesel Processor 5**
Ilesanmi Afolabi Daniyan and Khumbulani Mpofu

Chapter 3 **Application of Six Sigma in Semiconductor Manufacturing: A Case Study in Yield Improvement 27**
Prashant Reddy Gangidi

Chapter 4 **Design and Programming of a Wire Winder Device for Extrusion Activity in the Metal-Mechanical Industry 45**
Martha Roselia Contreras Valenzuela, Alejandro David Guzmán, Diana Lagunas, Gerardo Vera Dimas, Alina Martínez Oropeza, Viridiana León-Hernández, Alber Eduardo Duque Alvarez and Roy López Sesenes

Chapter 5 **Additive Manufacturing Applied to the Design of Small Satellite Structure for Space Debris Reduction 59**
Jonathan Becedas and Andrés Caparrós

Chapter 6 **Multiple Criteria Evaluation of Assembling Buildings from Steel Frame Structures 77**
Ruta Miniotaite

Preface

The manufacturing of industrial products comprises the manufacturing processes of component parts, the assembly process of individual parts in the assembled state, and mounting the finite product right at the end, this product being required to fulfill all the functions for which it has been designed.

The quality of the products and their competitiveness depend on the quality of the component parts, the efficiency of the manufacturing processes, and the right solutions for the assembly process of a newly developed product in close connection with the efficiency of the assembling technology.

Designing competitive products must also take into consideration the technological design (Design for Manufacturing concept), as well as the assembly design (Design for Assembly concept), and functional aspects of the new developed product at the end. Design for Assembly (DFA) is the method of product design for ease of part assembly, which involves optimization of parts or subsystems.

Design for manufacturing and assembly (DFMA) is the method for process and cost optimization of subsystems. Using the principles of DFMA, some efficiencies in assembling and manufacturing are obtained and the assembly operations are minimized.

The book aims to present applicable research in the field of design, manufacturing, and assembly realized by researchers affiliated to well-known institutes. The book has a profound interdisciplinary character and is addressed to researchers, engineers, PhD students, graduate and undergraduate students, teachers, and other readers interested in the assembly applications. I am confident that readers will find interesting information and challenging topics of high academic and scientific level within this book.

The book presents case studies focused on new design for special parts using the principles of DFMA, strategies that minimize defects in design and manufacturing applications, special devices produced to replace human activity, multiple criteria analysis to evaluate engineering solutions, and the advantages of using the additive manufacturing technology to design the next generation of complex parts, in different engineering fields.

Dr.Eng. Ancuţa Păcurar
Lecturer
Technical University of Cluj-Napoca
Cluj-Napoca, Romania

Chapter 1

Introductory Chapter: Applications of Design for Manufacturing and Assembly

Ancuța Carmen Păcurar

Additional information is available at the end of the chapter

http://dx.doi.org/10.5772/

1. Introduction

The occurrence and development of new and highly productive technological procedures lead to the need of enhancing the efficiency of assembly procedures as well as of material and part manipulation processes (object of the so-called logistics).

The manufacturing of industrial products involves processing the component parts, as well as assembling the individual parts in subassemblies and then assembling the final product that must fulfill all the functions for which it has been designed (**Figure 1**).

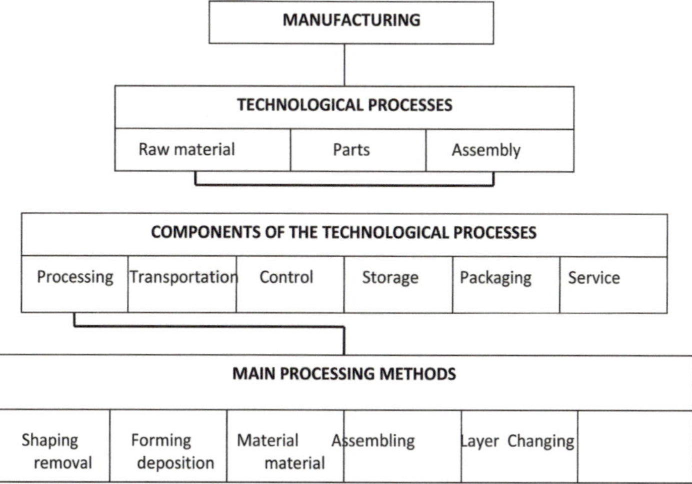

Figure 1. Place of the assembling in the manufacturing process.

Besides the functional characteristics of new products, the competitive design must take into account the principles of the design for manufacture (DFM), as well as the principles of the design for assembly (DFA).

The assembly process comprises the mounting procedures of tools, control devices, and jigs needed to perform the assembling operations. In view of establishing the optimal assembly method, as well as of its differentiation degree, the main factors that influence the assembly process must be analyzed (production batch, constructive characteristics of the product, and available manufacturing equipment).

In order to succeed, the researchers must design products in a right manner for optimal manufacturing, assembly, cost, quality, time, and functionality. Design for manufacturing and assembly (DFMA) is a methodology for product development or product improvement projects in which designers and manufacturing engineers work together instead of working separately. The two groups design the product's manufacturing and assembling processes; at the same time, they design the product itself. There are many CAD—integrated design for manufacturing programs—which help to identify and correct downstream issues in the design stage, leading to the reduction of the cycle time and the product development costs.

The main role of the mechanical engineer in the continuous industrial development is to be a creator by applying his/her knowledge and expertise to solve technical problems and to optimize the manufacturing processes.

The best performance in the field of mechanical manufacturing is obtained by "optimal operations" and "optimal processes." Therefore, the goal is to optimize the manufacturing procedures and their economic impact while making a product. This goal can be achieved by minimizing the manufacturing time. Within a company, certain types of cutting tools and machines characterized by certain technological parameters are used; therefore, to minimize the manufacturing time of a product, it is needed to optimize the machining method.

It is important to realize a design for manufacturing and assembly (DFMA) of the product which must provide an easy machining and assembling and also decrease manufacturing costs without having a negative impact on the quality requirements. It is important to know the fact that about 70% of the manufacturing costs of a product are determined by design decisions [1].

While product specification and costumer's requirements are important, there are other design issues to consider. The principles of design for manufacturing (DFM) are applied throughout the design process, and these affect all aspects from the design phase to the production phase. By applying DFM methods, the manufacturing costs can be estimated, while the costs of components and supporting production can be reduced.

The design for manufacturing system is a group of design principles that are structured to help the designer to reduce the costs and the manufacturing difficulties. The following rules can be mentioned: use of standard materials and components, standardized design, liberal tolerances, materials that are easy to process, avoidance of secondary operations, minimization of the manufacturing operations, and elimination of the intricate shapes [2].

The main principles that must be taken into consideration when designing for manufacturing are the selection of the adequate material; designed shape which ensures the selection of performance cutting tools with low cost; and selection of suitable tolerances for product design, which determine the type of the manufacturing. It is necessary to take into account the mechanical properties of the raw material, to be able to select the right material for the considered product. Due to the fact that the manufacturing time has a significant influence on the final cost of the part, cutting tools must be selected with the appropriate stiffness and strength, which provides a processing time as low as possible. The manufacturing cost is also influenced by the cost of the cutting tools. Therefore, the product must be designed to avoid using specialized, complex cutting tools [3].

By using these principles together with computer-aided design (CAD), computer-aided manufacturing (CAM), and finite element analysis (FEA), designers must find alternatives for the considered project, to obtain a design which assures a minimal machining time and a low-cost manufacturing.

Design for assembly (DFA) is the method of product design for ease of part assembly which involves optimization of parts or subsystems. The use of DFA allows a considerable reduction of the manufacturing costs, assembly time, and number of component parts, as well as increasing the productivity. Under such circumstances, it is possible to establish optimal assembly methods. Of course, the innovative contribution of experts is needed when using DFA software. When establishing the sequence of assembly procedures and selecting the appropriate equipment, the software users must use their expertise.

The main characteristics of DFA are listed below:

1. *Reducing the number of parts by combining or removing some of them*
2. *Improving the assembling accessibility*
3. *Improving the flexibility of components*
4. *Increasing the number of symmetric components*
5. *Optimizing the maneuverability of components (Whenever possible, stiff components are preferred instead of the flexible ones, because their manipulation is easier. Adequate surfaces are designed for mechanical gripping.)*
6. *Avoiding separate gripping elements*
7. *Providing parts having integral self-locking characteristics*
8. *Leading to modular design (Usage of standardized modules with the same functional requirements and standard interfaces that facilitate the interchange capabilities. Such an approach facilitates many options, quick design updates, as well as easier testing and servicing.)*

Author details

Ancuţa Carmen Păcurar

Address all correspondence to: ancuta.costea@tcm.utcluj.ro

Technical University of Cluj-Napoca, Cluj-Napoca, Romania

References

[1] http://www.unm.edu/~bgreen/ME101/dfm.pdf

[2] http://www.slideshare.net/NIELITA/dfm-design-principles

[3] https://www.mfg.com/sites/default/files/files/DFM%20CNC%20Machining.pdf

Chapter 2

Application of Design for Manufacturing and Assembly: Development of a Multifeedstock Biodiesel Processor

Ilesanmi Afolabi Daniyan and Khumbulani Mpofu

Additional information is available at the end of the chapter

http://dx.doi.org/10.5772/intechopen.80085

Abstract

Design for manufacturing and assembly (DFMA) is the method for process and cost optimization of subsystems, whole system as well as the entire manufacturing process. While minimizing assembly operations, it helps in eliminating component redundancy, facilitates assembly and manufacturing of products that are cost effective in terms of material requirements, parts production, labor and overhead. In this study, a multi feedstock biodiesel processor with intelligent systems for control and monitoring was developed using the principles of design for manufacturing and assembly. It consists of a 2 kW variable speed sparkless electric motor, reaction chamber, a thermostatically controlled 3 kW electric heater, saw dust insulation, ball valve, thermostat, funnel, a stirrer of diameter 20 mm with five blades that rotate in the reaction chamber and two baffles to control the splashing. The stirrer is driven by the electric motor. A 2 mm thick galvanized steel was used in the fabrication of the reaction chamber because of its high resistance to corrosion. This work provides a design framework for both small and large scale biodiesel plant for industrial, laboratory and experimental purposes. In addition, the assembly operations of the processor's components via the principles of DFMA were simplified to reduce ambiguity and redundancy. Hence, the overall processor is cost effective in terms of material requirements, parts production, labor and overhead.

Keywords: design, biodiesel, manufacturing and assembly, multi feedstock, cost optimization

1. Introduction

Design for manufacturing and assembly is a combined method developed to evaluate product assembly in order to enhance simplicity and cost effectiveness in manufacturing during

the product development. It is a two part analysis of product development namely; design for assembly (DFA) and design for manufacturing (DFM), with the aim to increase flexibility and profitability [1, 2], reduce overall cost in terms of material, production, labor and overhead [3, 4], as well as to reduce the complexity of manufacturing process [5]. A complex process increases manufacturing time, cost and rigidity. Design for assembly (DFA) is the method of product design for ease of part assembly which involves optimization of part or subsystem. According to Strienstra [6], DFA is a tool used in the design of product parts or subsystems that will transit into production at an effective cost, focusing on the number of parts, handling and ease of assembly. The aim is to minimize part lists without sacrificing quality, reduce assembly cost and to select the optimum cost of material that will be employed in manufacturing. On the other hand, design for manufacturing (DFM) is a method of design for ease of manufacturing via optimization of the process of product integration into the final product [7]. The process of optimization presents several optimum solutions for the manufacturing process hence DFM select the best solution out of the possible solutions. This reduces overall production cost as well as the complexity of the process of product development without compromising the standard. The process of product development starts with the conceptual design stage to design for assembly from where it transits to design for manufacturing and detailed design of the product. Depending on the orientation and geometry of different component parts, manufacturing processes are utilized as a single approach or in combination. This may include forming or machining processes, etc. Besides designing a component to function and fit for the intended application, it is important to put manufacturing processes into consideration. This proactive step is known as Design for Manufacture and Assembly (DFMA). It reduces the product development cycle while helping in ensuring that rework and time wasting activities are eliminated as parts are produced in the most judicious and economic way. According to [2, 8] other advantages of DFMA includes; correct interpretation of design information, optimization of design and manufacturing methods into one single step, increase in productivity with attendant decrease in production cost as well as high degree of precision in design and manufacturing. The basic principle for DFA involves the minimization of part list via the design for parts with self-locating and fastening ability. This makes the design simple as the use of locators and fasteners are reduced. In addition, parts are designed for ease of handling, insertion and retrieval. This increases design flexibility and minimizes the re-orientation of parts during assembly. Modular design of parts and part assembly from top to bottom is an effective DFA concept that can be employed to optimize the design phase. According to Whitney [9], product assembly can be done via manual, fully automated or semi-automated means. While manual assembly relies on human effort for assembly operations, automatic assembly uses robotic technology and computer controlled systems [3]. Semi-automatic relies partly on both human effort and intelligent systems for assembly operations. The method of assembly however depends on the nature of product, product complexity, environment, manufacturing structure, the efficiency required amongst other factors. Manual method of assembly has the merit of flexibility, cost effectiveness and mostly suitable for simple products with short assembly cycle. The demerits however lies in the low speed of assembly which increases assembly time, poor handling, and lower efficiency due to stress and fatigue on the part of personnel when compared to automatic means and most times not suitable for assembly of complex geometry.

Robots are programmable machines which receives signals from the system and environment to carry out programmed activities autonomously or semi autonomously [10]. They take decisions and interact with other interface as well as the central control system via the sensors and actuators. Robots combine the techniques of numerical control and remote control to replace human personnel with numerically controlled mechanical actuators. Robots are classified into two categories; artificially intelligent robots and non-artificially intelligent robots. Artificially intelligent robots are robots which are controlled by artificial intelligent programs involving learning, perception, problem-solving, language-understanding and logical reasoning to perform complex tasks while non-artificially intelligent robots simply carry out a defined sequence of instructions without the use of artificial intelligent programs. Robots possesses good material handling ability and suitable for assembly of complex products at higher efficiency and reduced assembly time. The use of robots for assembly is costly and some robots are not flexible enough when compared to manual method.

The emerging technology in product assembly involves the use of machines operated and controlled by computer programs. There are limitations in the rate of production using manual or robotic means because time is wasted during assembly. The computer controlled system uses some codes of instruction consisting of numbers, letters of the alphabets and symbols. These instructions are converted into electrical pulses which the machine's controls follow to carry out the assembly operations and the process is automated with the help of micro-controllers. They are costly, but highly efficient and can carry out any type of assembly operation with high degree of precision and accuracy within a short time. The pace of assembly delivery with high precision will offset the high initial cost. This work applies the design principles for design for manufacturing and assembly in the development of a multi-feedstock biodiesel processor limited to the production of 50 L of biodiesel per day. Several researchers have worked on the development of a biodiesel processor. For instance, Leevijit et al. [11] developed a continuous stirred tank reactor for the transesterification of palm oil to biodiesel. Process simulation was performed to optimize the mixing performance of the continuous reactor and the required time for the transesterification of palm oil in the optimized reactor was predicted as 5 h. However, the analysis of plant design was not sufficiently highlighted. Marjanovic et al. [12] also developed a batch reactor for the production of biodiesel from divers' feedstock but the analysis of integration of plant components was a missing link. Bello et al. [13] developed a batch biodiesel processor that can transesterify used and unused oil to biodiesel but there are some redundant complex geometries which increases manufacturing time, cost and rigidity. Musa [14] designed and constructed a pilot plant for the production of biodiesel from cottonseed. The capacity of the plant is limited to 20 L of biodiesel per day. The biodiesel pilot plant consists of a transesterification reactor with heater, a stirrer, chemical mixing tank, three glycerol settling tanks, and washing tank. However, the analysis of control circuitry and effect of process parameters were not studied. Bhachu et al. [15] designed a mobile biodiesel production plant, which is capable of producing 3000 L of biodiesel per week. The limitation lies in the fact that there are many redundant parts that increased the overall cost of manufacturing. Also, Highina et al. [16] reported on biodiesel production using *Jatropha curcas* oil in a batch reactor in the presence of zinc oxide as catalyst. The reactor is limited to single feed and use of heterogeneous catalyst only. Although, heterogeneous catalysts boast of several advantages over homogeneous catalysts in the conversion of feedstock to biodiesel [17–20]. The analysis of material handling was however not sufficiently highlighted. Hashem et al. [21] developed a bench top automated system that simulates an existing processor design with modifications necessary to make it run by computer control. The processor is adaptable to test alternatives and novel processing techniques in the laboratory on the bench top scale. The automation was done using a program controller integrated data acquisition software and Labview software to monitor and control the operation. The prototype developed is useful in simulating full scale automated system and in testing new processing technologies. The specific goals of the design are to minimize manual labor and cost while ensuring plant flexibility in processing multi feedstock. The aim of the work is to develop a multi-feedstock biodiesel processor with a capacity of 50 L per day using the principles of design for manufacturing and assembly (DFMA).

The objectives of the work are to:

i. design a smart multi-feedstock biodiesel processor with a capacity of 50 L per day with reduction in the part list;

ii. consider existing processors and reduce the overall manufacturing cost via redesign and elimination of redundant parts;

iii. increase the versatility and flexibility of the processor.

2. Methodology

This work uses the principles of design for manufacturing and assembly for development of a smart biodiesel processor. The design considerations are as follow;

i. The number of parts and number of interfaces which determines the simplicity or complexity of the overall assembly. The parts are listed in order of assembly and are assigned numbers to identify the parts. The complexity factor is expressed as Eq. (1).

$$CF = \left(\Sigma N_p \times \Sigma N_i\right)^{0.5} \tag{1}$$

where ΣN_p is the total number of part list and ΣN_i is the total number of part-to-part interface.

ii. The practical and theoretical minimum parts as well standard parts with their cost. Parts of different materials are identified and functional analysis is carried out with respect to existing design and possibility of redesign. This is a crucial consideration because the aim of product development is for profitability and to come up with reliable products that will meet customers' requirement by performing satisfactorily in service. Hence, the right balance must be struck between cost and component parts without sacrificing quality.

The theoretical part list efficiency is expressed by Eq. (2).

$$\frac{T_m}{T_n} \times 100\% \tag{2}$$

where T_m is the theoretical minimum number of parts and T_n is the total number of parts.

iii. Nature of the parts to be assembled. This determines the method of handling, insertion an alignment.

http://dx.doi.org/10.5772/intechopen.80085

iv. The method of joining the different parts together. This is followed by evaluation of the assembly and readjustment.

v. The determination of the right drafting and modeling tools in transforming conceptualized design into reality with parts specifications.

2.1. Design novelties

The design novelties of the developed processor are as follow:

a. Significant reduction in part list and short assembly cycle with parts designed for ease of handling, insertion and retrieval.

b. Provision of a smart processor that is flexible and versatile enough to process biodiesel from different feedstock via incorporation of all feedstock requirements for the production of biodiesel.

c. Provision of suitable template for scaling its development; and

d. Incorporation of intelligent systems for control and monitoring.

The intelligent systems for control and monitoring comprises of the following;

2.1.1. Arduino Uno microcontroller

The Arduino Uno is a microcontroller board that provides a simple and modular way of interfacing the real world with the computer to handle basic processing tasks on a chip while working with hardware sensors (**Figure 1**). The Arduino Uno uses the ATmega328 chip that supports 14 digital pins that can be configured as either input or output and 6 analog inputs [22].

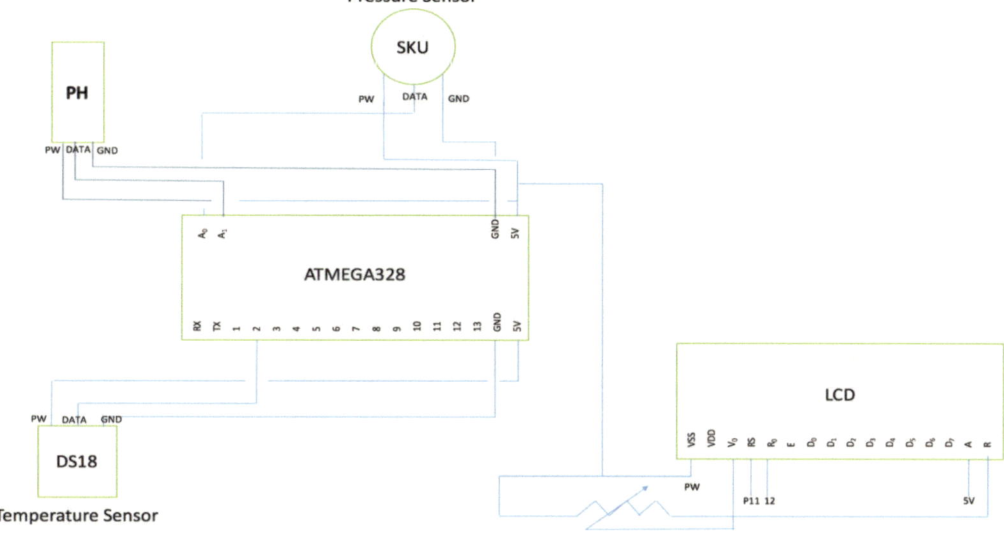

Figure 1. The monitoring system circuit with Arduino Uno microcontroller.

The technical specification of the Arduino Uno is presented in **Table 1**.

The Arduino is programmed using the Arduino IDE with source code will be written in C.

S/N	Item	Value	Remarks
1	Micro-controller	8-bit Atmel ATmega328p	1 mm sheet metal
2	Operational voltage	5 V	Input range: 7–12 V
3	Digital GPIO	14	6 capable of PWM
4	Analog IO	6	10-bit
5	Program memory	Flash 32 kb, EEPROM 1 kb	SRAM 2 kb
6	Clock speed	16 MHz	
7	USB	Type B socket	
8	Programmer	In-system firmware	USB-based
9	Serial communications	SPI, I2C	Software UART
10	Other	RTC, watchdog, interrupts	

Table 1. Technical specifications of the Arduino Uno.

2.1.2. Pressure transducer sensor

This measures the pressure of gas with a carbon steel alloy sensor material (**Figure 2**). It has a working pressure range of 0–1.2 MPa. The normal working temperature range is 0–85°C and the response time is approximately 2 ms.

Figure 2. Pressure transducer sensor.

It consists of an elastic material that deforms under the application of pressure and an electrical element which detects the deformation and transmits it as changes in voltage.

2.1.3. *pH meter kit*

This is a kit that measures pH of a substance. It is specially designed for the Arduino and has an accuracy of ±0.1 pH (at 25°C). The kit has a range of 0–14 pH. The kit consists of a pH sensor probe, a BNC connector and a pH 2.0 interface. The pH meter kit is shown in **Figure 3**.

Figure 3. pH meter kit.

2.1.4. *Temperature sensor*

This takes temperature readings for the plant to aid process insights. It has a temperature range of 0–400°C.

2.2. Processor design

The processor design involves the determination of a detailed, industrial-scale engineering model for a biodiesel processor and the associated parametric specification. The components parts are linked up with the aim of integrating and optimizing the entire process in the conversion of oil to biodiesel. The design involves the determination of the reactor geometry and capacity as well as the choice of construction material. The first design consideration is the reactor capacity which is determined by the mass balance of input and output streams. The sum of the volume of all streams entering the reactor gives the total liquid volume of the reactor for one batch. In order to meet an annual production of 18,000 L, daily production is calculated as 50 L per day.

2.3. Materials

The processor part list as well as the materials employed in the construction of a biodiesel processor is presented in **Table 2**.

S/N	Description	Quantity	Remarks
1	Sparkless electric motor (E2A), 3000 Hz, max 700 NCm	1	2 kW capacity
2	Reactor	1	2 mm thick galvanized sheet metal
3	Heater	1	3 kW of heat capacity of 0.44 kJ/kgK
4	Valve	1	regulatory valve
5	Insulator		Saw dust
6	Thermostat	1	TED-2001 (0–400°C)
7	Stirrer	1	Stainless steel
8	Funnel	1	Glass
9	Arduino Uno	1	Microcontroller
10	pH meter kit	1	43 × 32 mm
11	Pressure transducer sensor	1	¼″ 1.2 MPa
12	Temperature sensor	1	
13	Valves and fitting	2	½″ ball valve
			½″ adapter
			¾″ socket
			¾″ × ½″ bushing
			¼″ gas outlet valve
			½″ T-fitting
			½″ PVC pipe

Table 2. Processor parts list and materials employed.

2.3.1. Sparkless electric motor

This is a device powered by electric current to produce continuous motion. It rotates at a maximum speed of 300 rpm and attached to it, is the shaft. The power of the electric motor is rated as 2 kW and it is such that it is a variable speed electric motor from which different stir speed can be selected. It has a maximum frequency of 3000 Hz. The electric motor drives the stirrer which in turn stirred the raw materials inside the reaction chamber tank.

According to [23] a 2 kW electric motor is adequate to stir fluid of density 800–900 kg/m^3 at a speed of 300 rpm. This informs the choice of this type of electric motor.

2.3.2. Reactor

This is the main chamber where transesterification takes place. The reactor is pressurized and its chamber is constructed into cylindrical shape with galvanized steel. The reaction tank has a galvanized steel lid welded on. The lid for the tank is important in order to prevent escape of methoxide fumes during the process and also allows for distilling out the methanol after each batch. The lid also helps to prevent exposure to poisonous fumes, dangerous chemicals and fire. The capacity of the reaction tank is limited to 25 L per batch of vegetable oil.

2.3.3. Heater

The bottom of the cylindrical reaction chamber is fitted with an internal heating element. The heater is a 3 kW, 120 V a/c stainless steel heating element. It is used to raise the temperature of the reaction chamber and its content between 0 and 400°C. The processor chamber is heated by this thermostatically controlled electric heater and the temperature of the content is measured and regulated by means of a thermostat.

The fluid density is expressed as Eq. (3).

$$\rho = \frac{m}{v} \tag{3}$$

where v is the volume (0.027 m^3) and ρ is the fluid density (830 kg/m^3), then the mass m of the fluid is calculated 22.41 kg.

The weight of the fluid is expressed as Eq. (4).

$$w = mg \tag{4}$$

where m is the mass of the fluid (22.41 kg) and g is the acceleration due to gravity (9.81 m/s^2).

Therefore, the weight of the fluid is calculated as 219.84 N.

The specific heat capacity of vegetable oil is 2340 J/kg K, weight (W) of 25 L of vegetable oil is 219.84 N, the heater has a power rating of 3 kW, assuming the ambient temperature of 30°C, and the mixture is heated to a temperature of 60°C, therefore, the heat transferred to the fluid (workdone by the heater) is given as Eq. (5).

$$Q = mc\Delta T \tag{5}$$

where m is the fluid mass (kg); c is the specific heat capacity of the fluid (J/kg K); and ΔT is the temperature difference between the final T_2 and ambient temperature T_1.

Hence, 1573 kJ is the quantity of heat transferred to 25 L of vegetable oil per batch. The assumed electrical to heat conversion efficiency (η_e) is 80%. The heat loss coefficient for the reaction tank is 1.0 J/S-m^2-°C and the surface are of the reactor is 0.5430 m^3.

The power rating of the heater is expressed as Eq. (6).

$$P = \frac{W}{T} \tag{6}$$

where P is the power rating of the heater (3 kW); W is the workdone by the heater (1573 kJ).

Therefore, the estimated heating time for the used vegetable oil is 530 s.

The rate of heat loss is expressed as Eq. (7).

$$R_L = \alpha * SA * \Delta T \tag{7}$$

where α is heat loss coefficient for the reaction tank (1.0 J/S-m^2-°C); SA is the total surface area of the reactor (0.5430 m^3); ΔT is the temperature difference between the final T_2 and ambient temperature T_1 (30°C).

Therefore the rate of heat loss R_L is calculated as 16.29 J/s.

2.3.4. Valve

The ball valve is fitted to the bottom of reaction tank in order to control the rate of discharge of reaction products from the reaction chamber.

2.3.5. Insulator

The reaction tank is well lagged all through to keep the temperature of the reaction tank and its content constant by preventing heat loss. The insulating material employed is saw dust and has a thermal resistivity of 33.3 mK/W. Insulated tanks maintain heat better than un-insulated ones due to continuous escape of heat from the tank.

2.3.6. Thermostat

This is essential for temperature regulation or control. It is a safety device capable of gauging temperature and helps in accurate temperature control as well as prevention of pressure build up within the reaction tank. The temperature range of the temperature regulator is between 0 and 400°C. This allows transesterification reaction to take place at selected temperatures. A temperature probe was attached to sense the temperature inside the reaction tank for two reasons; to indicate that the oil has been pre-heated so that methanol/catalyst can be loaded into the reaction tank and to keep the temperature of the solution between a given temperature range during the reaction loop. The outside temperature probe is to indicate when the inline heater should be engaged or disengaged.

2.3.7. Shaft

A shaft of 20 mm diameter is used, which ensures satisfactory strength and rigidity when the shaft is transmitting power under different operating and loading conditions. It is fabricated from stainless steel because of its high resistance to corrosive effects of chemicals and catalysts, since the shaft often comes in contact with chemicals and catalyst during mixing process at high temperatures.

2.3.7.1. Design of shaft

Shaft design consists primarily of the determination of the correct shaft diameter to ensure satisfactory strength and rigidity when the shaft is transmitting power under various operating and loading conditions [24].

Using the relationship given in Eq. (8).

$$\frac{T}{J} = \frac{\tau}{r} = \frac{G\theta}{l} \tag{8}$$

where G is the modulus of rigidity (N/m^2); L is the length of shaft (m); θ is the angle of twist (degree); T is the total resisting torque (N); τ is the maximum shearing stress (N/m^2); J is the polar moment of inertia (m/s^4); R is the radius of shaft in (m).

From the relation in Eq. (9),

$$\frac{T}{J} = \frac{\tau}{r} \tag{9}$$

The resisting torque is expressed as Eq. (10).

$$T = \frac{P}{2\pi N} \tag{10}$$

where P is the power rating of the electric motor (2 kW); N is the number of revolution per sec (400 rev/s).

Hence, the total resisting torque is calculated as 0.80 N.

Furthermore, the polar moment of inertia is expressed as Eq. (11).

$$J = \frac{\pi d4}{32} \tag{11}$$

where d is the diameter of the shaft (mm).

According to [25], using τ = 55 MN/m^2 (maximum shearing stress of steel from which the shaft is made) and substituting for T in Eq. (8), the shaft diameter is calculated as 17 mm and designed as 20 mm to a safety factor of 1.2. The weight of the shaft is negligible since the shaft is vertical in orientation for mixing, it is subjected majorly to twisting moment.

Stainless steel was employed for the fabrication of the shaft and galvanized steel for the reaction tank in order to overcome the problem of corrosion.

2.3.7.2. *Design for shaft deflection*

For a plate of length d (mm), shorter side b and thickness t (mm), the maximum deflection y_{max} (Eq. (12)) is found to occur at the centre of the plate.

$$\mathrm{y}_{\mathrm{max}} = \frac{\alpha q b^4}{E t^3} \tag{12}$$

where q is the fluid weight and the value of factor α depends on the ratio of d/b and E is the modulus of elasticity (N/m^2).

The maximum bending moment also occur at the centre of the plate and are given by the relationships expressed by Eqs. (13) and (14).

$$\mathrm{M}_{\mathrm{XYmax}} = \beta_1 q b^2 \tag{13}$$

$$\mathrm{M}_{\mathrm{YXmax}} = \beta_2 q b^2 \tag{14}$$

The factors β_1 and β_2 are given for an assumed value of Poisson's ratio v equal to 0.3.

Furthermore, recall the weight of the fluid given by Eq. (4).

Then the weight of the fluid is 219.84 N.

The maximum deflection given by Eq. (12) is calculated where q = 219.84 N; b = 0.6 m; $t = 2.0 \times 10^{-3}$ m; E = 200 GN/m^2; α = 0.0843.

Hence, maximum deflection $\mathrm{y}_{\mathrm{max}} = 1.5 \times 10^{-3}$ mm. This small value of the maximum deflection indicates that the processor has satisfactory strength to withstand the loading forces without significant distortion.

2.3.8. Funnel

After completion of the reaction, the product is transferred into a separating funnel for a certain time interval (approximately 24 h) for phase separation. Since the solubility of methyl ester is low, the glycerine tends to collect at the bottom. With excess alcohol, the unconverted triglycerides should essentially be zero. However, some monoglyceride and diglycerides must be present [26]. Due to their polarity, partially reacted glycerides should be preferentially attracted to the glycerine phase and then removed when the phase is separated. The funnel is employed during separation of biodiesel from the glycerine. It is also used for separating washed methyl esters from impurities such as water, sodium hydroxide and glycerine.

2.4. Construction of the biodiesel processor

A 2 mm thick galvanized steel was used in the fabrication of the reaction tank because of the following reasons:

- **i.** It does not catalyze the oil unlike copper;
- **ii.** Its high resistance to pressure and temperature;
- **iii.** Its ability to withstand the actions of chemicals and catalyst without any sign of rust or corrosion.

2.5. Design for volume

The volume of the reaction tank (cylinder) is given by Eq. (15).

$$V = \pi r^2 h \tag{15}$$

where $\pi = 22/7$; r is the radius of the cylinder (0.12 m) and h is the height of cylinder (0.60 m).

The capacity of the processor is limited to 25 L of vegetable oil per batch.

2.6. Total surface area of the processor

The total surface area of the reactor is expressed by Eq. (16).

$$\text{T.S.A} = 2\pi r^2 + 2\pi rh \tag{16}$$

where $\pi = 22/7$ and R is the radius of the cylinder (0.12 m).

Then the total surface area of the tank is 0.5430 m^3.

2.7. Design for manufacturing and assembly

Existing biodiesel processor has a separate pre-treatment tank either as a stand-alone facility or in addition to the reactor which makes the overall assembly expensive and cumbersome. The application of the design for manufacturing and assembly (DFMA) in the development of the biodiesel processor eliminates the need of the pre-treatment tank while the reactor serves both functions. Depending on the nature of feedstock, the incorporation of intelligent systems for control and monitoring helps in determining the need for the pre-treatment process or otherwise. In addition existing biodiesel processor has a base which makes drainage of the product a challenge. The application of DFMA in the development of the biodiesel processor replaces the base with a conical base which makes drainage quite easy. Other significant improvements in the new design compared to the existing design via the application of the DFA are discussed in Section 2.9. Autodesk inventor was employed in the design and assembly drawing of the processor the developed processor.

Figures 4–9 show the various views of the assembly diagram of the existing processor while **Figure 10** shows the new design for the biodiesel processor upon the application of design for manufacturing and assembly.

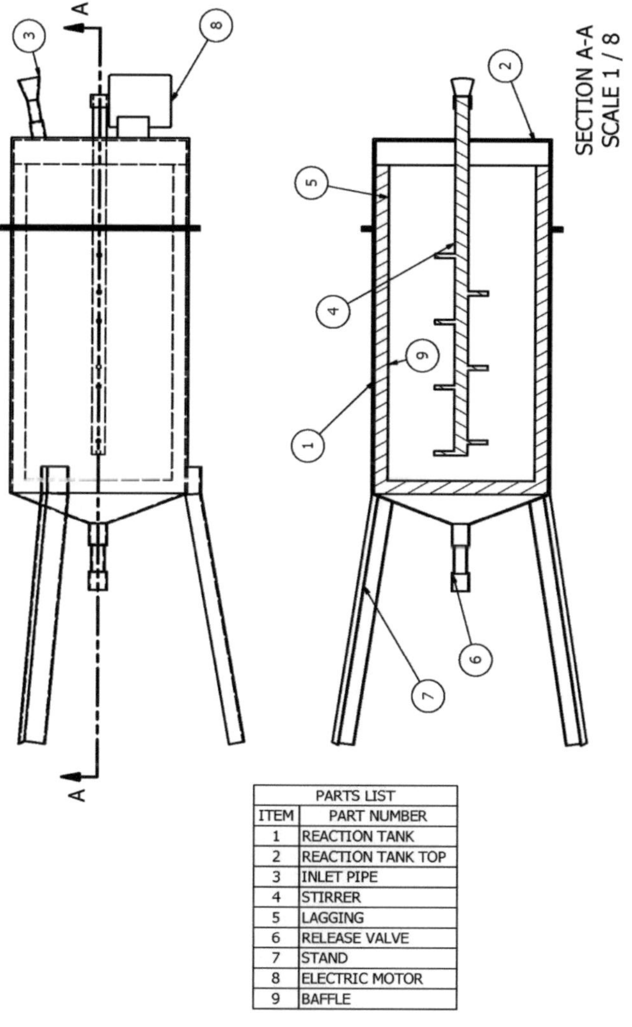

PARTS LIST	
ITEM	PART NUMBER
1	REACTION TANK
2	REACTION TANK TOP
3	INLET PIPE
4	STIRRER
5	LAGGING
6	RELEASE VALVE
7	STAND
8	ELECTRIC MOTOR
9	BAFFLE

Figure 4. Sectional view of the developed processor.

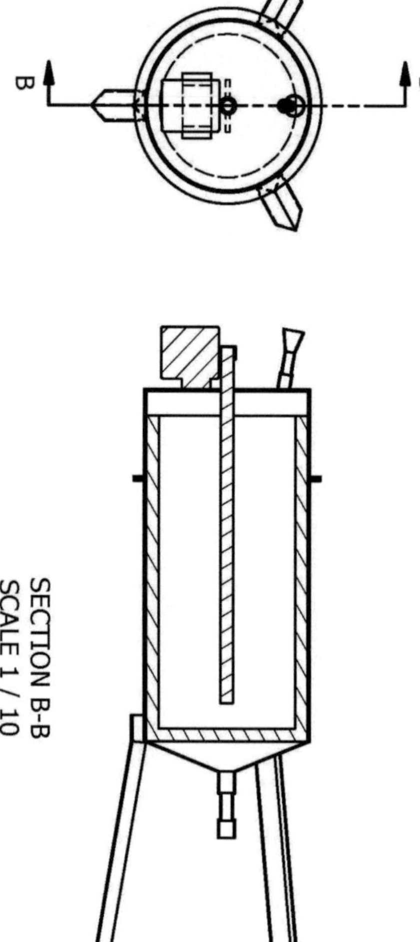

Figure 5. Sectional view (top) of the developed processor.

2.8. Method of assembly

The procedural steps for the assembly operation are as follow;

- **i.** Rolling of the inner cylinder of diameter 240 mm and height and height 350 mm.
- **ii.** Rolling of the outer cylinder of diameter 360 mm and height 500 mm.
- **iii.** Lagging of the space in between the inner and outer cylinder followed by welding of the inner and outer diameter to give a single cylinder of diameter 360 mm with an overall height of 600 mm.
- **iv.** Welding of the stand of height 180 mm to the cylinder.

v. Welding of five blades to the shaft of diameter 20 mm and length 400 mm to form the stirrer assembly.

vi. Welding of the flat cover to the tank and use of hinges to create a top opening for maintenance purposes.

vii. Attachment of the electric motor to the stirrer assembly using bolt and nuts.

viii. Finishing operations: deburring and painting.

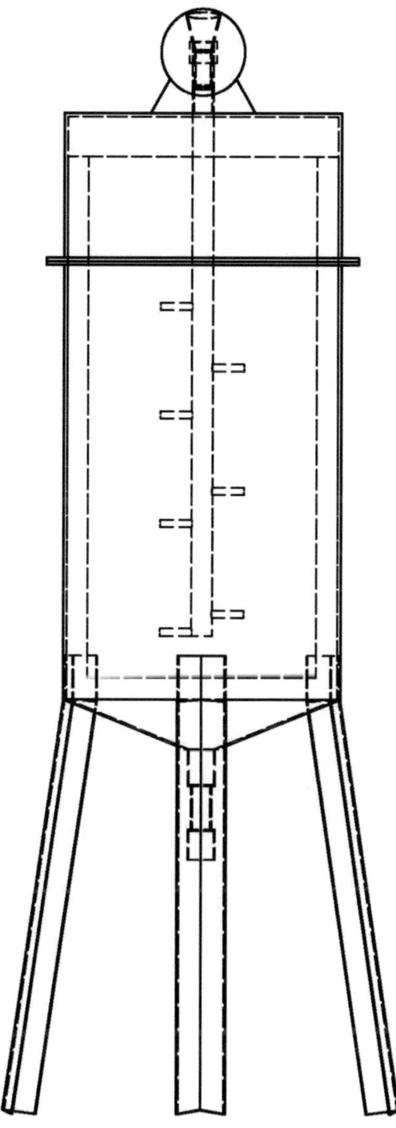

Figure 6. Right view of the developed processor.

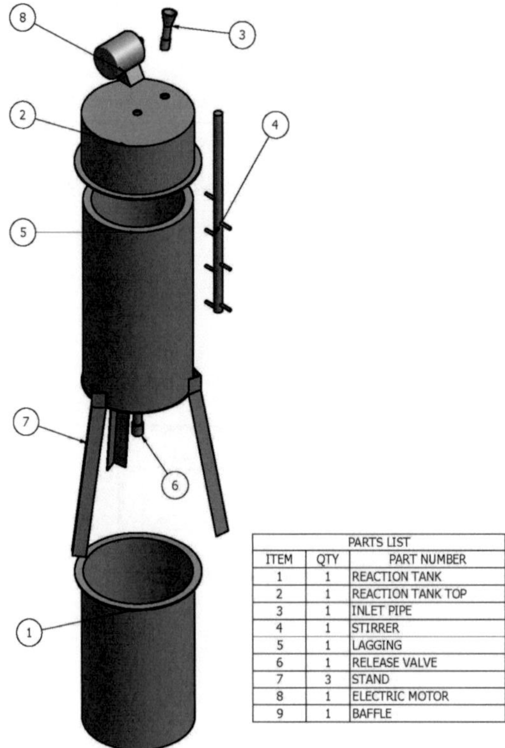

PARTS LIST		
ITEM	QTY	PART NUMBER
1	1	REACTION TANK
2	1	REACTION TANK TOP
3	1	INLET PIPE
4	1	STIRRER
5	1	LAGGING
6	1	RELEASE VALVE
7	3	STAND
8	1	ELECTRIC MOTOR
9	1	BAFFLE

Figure 7. Exploded view of the developed processor.

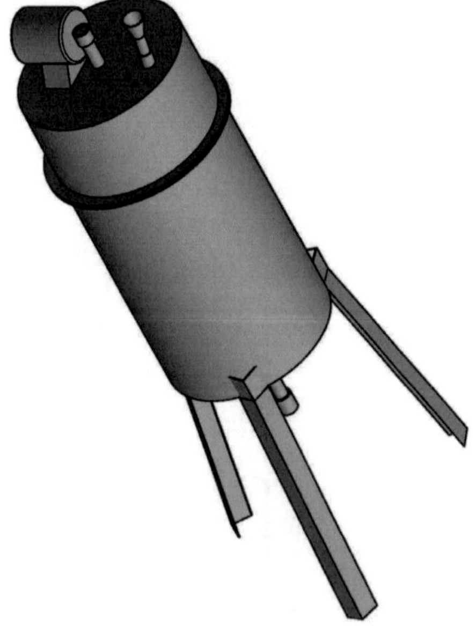

Figure 8. Isometric view of the developed processor.

http://dx.doi.org/10.5772/intechopen.80085

Figure 9. The developed biodiesel processor.

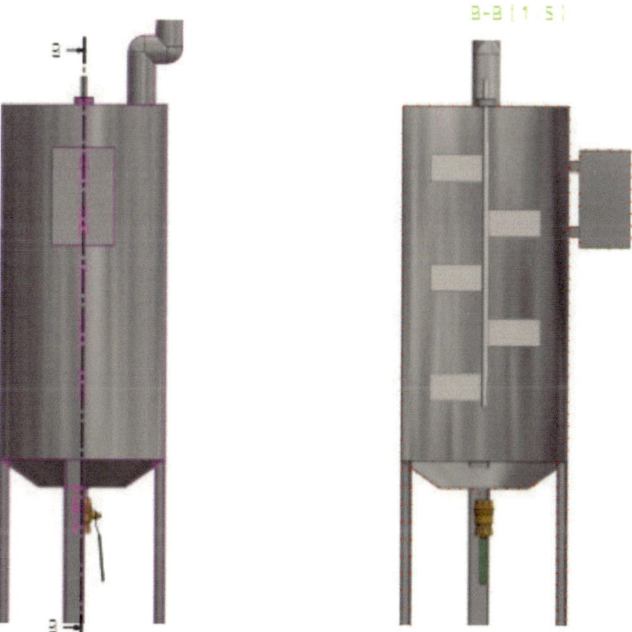

Figure 10. The new design for the biodiesel processor.

2.9. Differences between the existing and new design

Following the application of the design for manufacturing and assembly (DFA) in the development of the biodiesel processor, the following are the differences between the existing and new design;

i. Existing design involves rolling of the cylinder cover whose inner diameter is 240 mm and outer diameter 360 mm. The space in between the inner and outer cylindrical cover was also lagged after which the two are brought together by welding. In the new design, the cylindrical cover replaced with a flat cover welded to the tank with an opening at the top for maintenance purposes.

ii. Existing design has seven blades welded to the shaft to form the stirrer assembly with four baffles to prevent splashing. In the new design fives blades ensures homogeneity of the mixture with two baffles to prevent splashing of the mixture.

iii. In the existing design, the shaft was joined to the cylindrical cover using bearing, bolt and nuts while this was simplified in the new design with the replacement of the cylindrical cover with a flat cover. There was no joining of the shaft to the cover. This makes disassembly quite easy for maintenance purposes. The shaft was passed through the flat cover with the electric motor attached to the top of the shaft using bearing, bolts and nuts.

iv. Welding of the rectangular control panel whose length is 300 mm and breadth 200 mm, as well as welding of the stand of height 550 mm to the control panel was done in the existing design. In the new design, the fabrication of the control panel was replaced with a smart control panel bought out at price lesser than the cost of fabrication in the existing design.

v. Smart and intelligent systems were incorporated into the new design for process monitoring and control which was lacking in existing design.

vi. The application of the design for manufacturing and assembly (DFA) in the development of the biodiesel processor replaces the base with a conical base which makes drainage quite easy.

vii. The new design is smart, robust and more efficient than the existing design. The cost comparison of the new and existing design is shown in **Tables 3** and **4**, respectively

2.10. Summary of development analysis

The bill of engineering measurements (development and evaluation) is presented in **Table 5**.

S/N	Parts	Process	Materials	Cost (USD)
1.	7 support legs	Cutting/welding	Mild steel	77
2.	Cylindrical body	Cutting/rolling	Galvanized steel	30
3.	Cylindrical cover	Cutting/rolling	Galvanized steel	60
4.	Control panel	Cutting/welding	Galvanized steel	60
5.	Stirrer assembly	Cutting/ welding	Stainless steel	75
			Total	337

Table 3. Cost analysis of the existing design.

S/N	Parts	Process	Materials	Cost (USD)
1.	3 support legs	Cutting/welding	Mild steel	33
2.	Cylindrical body	Cutting/rolling	Galvanized steel	70
3.	Flat cover	Cutting/welding	Galvanized steel	12
4.	Control panel	Bought out	Galvanized steel	30
5.	Stirrer assembly	Cutting/welding	Stainless steel	65
			Total	210

Table 4. Cost analysis of the new design.

Item #	Name	Quantity	Description	Unit price (USD)	Total price (USD)
1	Stainless steel rod	1	0.60 m length	60	60
2	Mild steel	1	2 m length	45	45
3	2 kW electric motor	1	AC motor with speed control	70	70
4	Ball valve	3	Outlet and inlet valves	0.55	1.64
5	Galvanized sheet metal	3 rolls	Square cross section	41	123
6	Welding electrodes	1 packet	E020, E6023	2.73	2.73
7	Heater	1		8.22	8.22
8	Temperature probe	1		8.22	8.22
9	Contactor	1		10.95	10.95
10	Funnel	1		8.22	8.22
11	Neon and switch light	2		2.73	2.73
12	Arduino Uno	1	Microcontroller	17.8	17.8
13	Pressure transducer sensor	1	¼" 1.2 MPa	14.79	14.79
14	Analog pH meter	1	43 × 32 mm	34.2	34.2
15	LCD	1		3.28	3.28
16	Electric cables	5 yards		1.40	6.84
17	Angle iron		1	20	20
	Control panel		1	30	30
18	Resistors and capacitors			2.75	2.75
19	Transport			40	40
20	Labor			12.3	12.3
21	Performance evaluation			275	275
				Subtotal	752.67
22	Miscellaneous		10% of total costs		75.267
	Total				**827.937**

Table 5. Bill of engineering measurement.

From **Tables 4** and **5**, the development of the biodiesel processor was cost effective with the elimination of pre-treatment and use of simple manufacturing processes.

2.11. Design specifications

The specifications in accordance with design is as follow:

i. Size	Fits in a 1.524 m × 0.609 m space
ii. Operating temperature	20–90°C
iii. Stir speed	50–600 rev/min
iv. Production amount	25 L per batch
v. Production time	3 h per batch
vi. Methanol consumption for unused oil	200 mL per 33 mL of oil olein vegetable oil
vii. Methanol consumption for used oil	200 mL per 25 mL of oil frying oil
viii. Catalyst consumption	4.5 g per 200 mL methanol

3. Contribution to knowledge

The work will contribute to knowledge as follows:

- **i.** Application of the principles of design for manufacture and assembly in the development of a smart biodiesel processor.
- **ii.** Improvement in process control and monitoring via the use of sensors and a micro-controller.
- **iii.** Incorporation of low-cost monitoring system in the biodiesel processor.
- **iv.** Provision of design framework for both small and large scale biodiesel processor for industrial, laboratory and experimental purposes.
- **v.** Development of a processor for processing biodiesel from different feedstock which could serve as a template for scaling its development.

4. Conclusions

This work brings about the development of a smart small scale biodiesel processor with a capacity of 50 L per day. The processor incorporate all feedstock requirements for the production of biodiesel hence, both alkali-catalyzed and acid-catalyzed transesterification can be undertaken with the processor. In addition, the assembly operations of the processor was simplified to reduce ambiguity and redundancy. This saves a total cost of 127 USD with increased efficiency and robustness. Hence, the overall processor is cost effective in terms of material requirements, parts production, labor and overhead. The principles of DFMA also minimizes part requirements without sacrificing quality. This reduces the overall assembly and manufacturing cost.

The development and performance evaluation of the processor cost a total sum of 827.937 USD. This is relatively cost effective considering the price of ready-made biodiesel processors in the market. Using the principles of DFMA, This work provides a design framework for both small and large scale biodiesel plant for industrial, laboratory and experimental purposes.

Author details

Ilesanmi Afolabi Daniyan* and Khumbulani Mpofu

*Address all correspondence to: afolabiilesanmi@yahoo.com

Faculty of Applied Sciences, Department of Chemical and Biological Sciences, Vancouver, Canada

References

[1] Boothroyd G. Early cost estimation in product design. Journal of Manufacturing Systems. 1994;**7**:183-192

[2] Naiju CD, Jayakrishnan V, Warrier V, Rijul B. Application of Design for Manufacturing and Assembly (DFMA) methodology in the steel furniture industry. In: National Conference on Technological Advancements in Engineering. 2016. pp. 1-4

[3] Boothroyd G, Dekker M. Assembly Automation and Product Design. New York; 1992. pp. 1-10

[4] Katja T, Miika J, Jari P. Activity based costing and process modelling for cost-conscious product design: A case study in a manufacturing company. International Journal of Production Economics. 2002;**79**(1):75-78

[5] Jain VK. Manufacturing Processes. India Institute of Technology, Kanpur, India: Department of Mechanical Engineering; 2015. pp. 1-34

[6] Strienstra D. Introduction to Design for Assembly and Manufacturing. 2012. pp. 1-81

[7] Lucchetta G, Bariani PF, Knight WA. Integrated design analysis for product simplification. CIRP Annals-Manufacturing Technology. 2009;**54**(1):147-150

[8] Ejaz S, Arjun MG, Naiju CD, Annamalai K. Conceptualizatio Design for Manufacture and Assembly (DFMA) of juice mixer grinder. In: National Conference on Advances in Mechanical Engineering. India: NCAME-2011; 2011. pp. 297-302

[9] Whitney D. Mechanical Assembly and its Role in Product Development. 2005. pp. 1-29

[10] O'Sullivan D. Industrial Automation. Portugal: Universidade do Minho; 2009. pp. 1-62

[11] Leevijit T, Wisutmethangoon W, Prateepchaikul G, Tongurai C, Allen M. Transesterification of palm oil in series of continuous stirred tank reactors. Asian Journal on Energy and Environment. 2006;**7**(3):336-346

[12] Marjanovic AV, Stamenkovic OS, Todorovic ZB, Lazic ML, Veljkovic VB. Kinetics of the base-catalysed sunflower oil ethanolysis. Fuels. 2010;**89**(3):669-671

[13] Bello EI, Daniyan IA, Akinola AO, Ogedengbe TI. Development of a biodiesel processor. Research Journal in Engineering and Applied Sciences. 2013;**2**(13):182-186

[14] Musa IA. Design & construction of a pilot plant for the production of biodiesel from cotton seed oil [M.Sc. thesis]. 2007. pp. 1-117

[15] Bhachu C, Chow N, Christensen A, Drew A, Ishkintana L, Lu J, Poon C, Villamayor C, Setiaputra A, Yan T. Design of a Portable Biodiesel Plant. University of Brititsh Colombia; 2005. pp. 1-137

[16] Highina BK, Bugaje IM, Umar B. Liquid biofuel as alternative transport fuel in Nigeria. International Journal of Petroleum Technology Development. 2012;**1**:1-15

[17] Olutoye MA, Wong SW, Chin LH, Asif M, Hameed BH. Synthesis of fatty acid methyl esters via transesterification of waste cooking oil by methanol with a barium-modified Montmorillonite K10 catalyst. Renewable Energy. 2016;**86**:392-398

[18] Tan YH, Abdullah MO, Hipolito CN, Taufiq-Yap YH. Waste ostrich and chicken-egg-shells as heterogeneous base catalyst for biodiesel production from used cooking oil: Catalyst characterization and biodiesel yield performance. Applied Energy. 2015;**2**:1-13

[19] Yusuff AS, Adeniyi OD, Olutoye MA, Akpan UG. Development and characterization of a composite anthill-chicken eggshell catalyst for biodiesel production from waste frying oil. International Journal of Technology. 2018;**1**:110-119

[20] Olutoye MA, Hameed BH. A highly active clay-based catalyst for the synthesis of fatty acid methyl ester from waste cooking palm oil. Applied Catalysis A: General. 2013;**450**:57-62

[21] Hashem D, Huang S, Pearson J, Farag IH. Biodiesel Processor Automation. Department of Chemical Engineering. United States: University of New Hampshire; 2012. pp. 1-12

[22] Goodwin S. Smart Home Automation with Linux and Raspberry Pi. 2nd ed. New York: A Press; 2013

[23] Srivastava A, Prasad R. Performance of a diesel generator fuelled with palm oil. Fuel Renewable Sustainable Revolution. 2000;**4**:105-140

[24] Hall AS, Holowenko AR, Laughlin HG. Machine Design. United States: Purdue University Press; 1980. pp. 113-130

[25] Ashby M, Shercliff H, Cebon D. Materials Engineering, Science, Processing and Design. India: Rajkamal Press; 2007. pp. 21-31

[26] Ma F, Clements LD, Hanna MA. The Effects of Catayst, Free fatty acids and Water on Transesterification of Beef tallow. Trans ASAE. 1998;**41**(5):1261-1264

Chapter 3

Application of Six Sigma in Semiconductor Manufacturing: A Case Study in Yield Improvement

Prashant Reddy Gangidi

Additional information is available at the end of the chapter

http://dx.doi.org/10.5772/intechopen.81058

Abstract

The purpose of this chapter is to outline systematic implementation of the Six Sigma DMAIC methodology as a case study in solving the problem of poor wafer yields in semiconductor manufacturing. The chapter also describes well-known industry standard business processes to be implemented and benchmarked in a semiconductor wafer fabrication facility to manage defect and yield issues while executing a Six Sigma project. The execution of Six Sigma enabled identification of the key process factors, root cause analysis, desired performance levels, and Cpk improvement opportunities. Implementing multilevel factorial design of experiments (DOE) study revealed critical input parameters on process tools contributing to defect formation. Improvement performed on these process tools resulted in in-line defect reduction and ultimately improving final yields.

Keywords: Six Sigma, semiconductor manufacturing, DMAIC, defects, yield, design of experiments (DOE)

1. Introduction

Six Sigma framework is a continuous improvement strategy that minimizes defects and process variation toward an achievement of 3.4 defects per million opportunities in design, manufacturing, and service-oriented industries [1–3]. Six Sigma practitioners often lead cross-functional teams in an organization to find and eliminate the causes of the errors, defects, lead, and cycle time delays in business processes. With rapid advancements in computers, artificial intelligence, and automotive vehicles, biomedical imaging semiconductor manufacturers find themselves constantly battling the demanding needs of the industry to sustain Moore's law and manufacture smaller chips to support next-generation software and hardware products.

In order to manufacture nanometer range scale chips, there is a tremendous impetus toward developing advanced process control and measurement system capabilities. Defects become a big challenge in the efforts to reduce feature size in most semiconductor fabs as they negatively impact product yields [4]. Six Sigma methodology is often neglected in most fabs, and this chapter gives an overview of its importance and how it can be implemented to reduce defects with the help of a general case study. This chapter presents the step-by-step application of the Six Sigma define, measure, analyze, improve, control (DMAIC) approach to eliminate defects in a lithography process of a semiconductor manufacturing organization. This has helped to reduce defects in the process and thereby improve the final probe yields on a critical technology node. During the measure and analyze phases of the project, data were collected from the processes to understand the baseline performance and for validation of causes. These data were studied through various graphical and statistical analyses. Chi-square test, ANOVA, design of experiments (DOE), control charts, fishbone analysis, FMEA, etc. were used to make meaningful and scientifically proven conclusions about the process and the related causes [5].

2. Six Sigma literature review

The primary methodology of Six Sigma is the application of DMAIC problem-solving steps. A brief explanation for each of the steps involved is as follows:

Define (D): The first stage focuses on analysis of customer identification, feedback, and requirements along with forming the project stakeholder team [6]. The project team looks at critical to quality (CTQ) and cost of quality (COQ) improvement projects that need to be addressed keeping the end goal of customer satisfaction in mind by defining project scope/problem statement and budget scheduling. It is very critical to define an accurate problem statement along with the scope to ensure the Six Sigma project team will invest all the time, skills, and resources in the right direction.

Measure (M): This is the data collection phase where types of data, measurement scales, and sampling and collection methods are evaluated. All the initial metrics of the business case are established to measure the problem scale [2, 7].

Analyze (A): Measure and model relationships between variables, hypothesis testing, root cause analysis using cause and effect analysis tools such as fishbone/Ishikawa diagrams, 8D methodology, and 5 Whys analysis [2, 7].

Improve (I): Post root cause analysis, this phase tries to understand optimum levels of factors responsible for causing the problem via design of experiments (DOE), giving insights to determine corrective and preventative actions. This stage involves lean strategies such as the Kaizen Blitz, poka-yoke (mistake-proofing), cycle time reduction, etc. [7–9].

Control (C): Primary objective in this last phase is to maintain and sustain control over the process and suggest improvement activities to minimize variation and defects. Statistical process control (SPC), total productive maintenance (TPM), and control plan development are some key tools used in the control phase [10].

3. Six Sigma application to defect reduction in semiconductor fabrication sites (FABS)

In this section, some business process areas have been identified to focus on benchmarking before considering Six Sigma project execution. Once these business process areas are well established, then only deploying Six Sigma teams would be beneficial to see tangible results. Motorola was one of the pioneers of Six Sigma methodology along with General Electric. Apart from Motorola, no other semiconductor manufacturing company has openly advocated the use of Six Sigma but has definitely inherited a lot of concepts and molded them into different terminologies. A big challenge to Six Sigma implementation is management support, and based on historical success rates of such projects initiated at their respective firms, the management decides to stick to their existing problem-solving methodologies or use some concepts from Six Sigma and other techniques used in industries such as aviation and automotive.

The most important goal for any semiconductor fab is to improve the final product yields [4]. Yield is directly correlated to contamination, design margin, process, and equipment errors along with fab operators [11]. **Figure 1** referenced from Integrated Circuit Engineering Corp. shows the ranking of top yield loss causing problems across various fabs [4]. Six Sigma DMAIC methodology can be used as an effective quality and reliability management tool to solve most of these issues, and several literature papers in the form of case studies have been published regarding the same.

Sources of random defects could be the equipment, fab personnel, process margins, process chemicals and gases, or cleanroom itself. Data collected from Integrated Circuit Engineering Corp. (ICE) over the last three decades have been shown in **Figure 2**.

Human and cleanroom sources of contamination have been steadily declining due to advanced training in this field being developed over the years at various universities and corporations along with rapid strides in automation and artificial intelligence that have modernized clean rooms and minimized human contact in handling wafers. Engineers and upper-level management of these fabs must adopt a systematic methodology in resolving yield losses that occur due to process and equipment variations, and in this chapter, some basic business processes

PROBE YIELD LOSS SOURCE	YIELD LOSS (%)	PERCENT OF TOTAL PROBE YIELD LOSS
Contamination	40	80
Design Margin	5	10
Process Variation	3	6
Lithography Errors	1	2
Material based defects	1	2
Total Loss	**50**	2
Probe Yield (100% - DIE LOSS) = 50 %		

Figure 1. Ranking of yield loss causing problems in fabs.

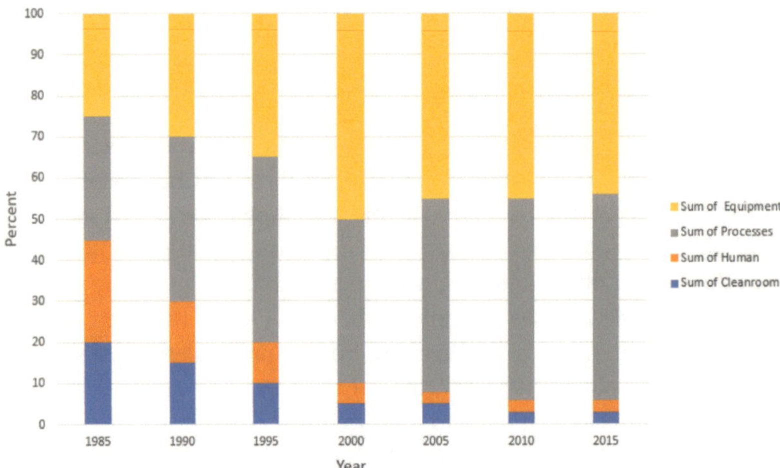

Figure 2. Bar chart showing wafer contamination source.

have been described that must exist in a fab for it to achieve maximum operational efficiency and produce high-quality chips. Six Sigma DMAIC methodology could then be used as and when required to improve these business processes.

3.1. Contamination and control protocols

Semiconductor processing involves several process fluids and gases, especially in lithography, film/metal deposition, etching, and cleaning steps. These fluids and gases contain impurity elements that can be dangerous to silicon devices [4]. These elements could be classified as the heavy metals, alkali metals, and light elements. Heavy metals such as Fe, Cu, Ni, Zn, Cr, Au, Hg, and Ag could result in wafer scraps and back end yield fallout due to corrosion in electroplating and metal deposition processing steps. Alkali metals such as Na, K, and Li and light elements like Al, Mg, Ca, C, S, Cl, and F could pose processing problems resulting in defects, which could be yield killing [4]. These elements also sometimes accumulate along the chambers, handlers, chucks, etc. of various equipment used in fabs resulting in tool downtime which greatly affects production schedules. Moreover, these elements could also pose safety problems to fab personnel. To understand potential risks and sources of these impure elements, a highly cross-functional FMEA team can be deployed to map out all the processes where source chemicals and gases are used along with identifying potential fail modes, severity, occurrence, and detection capabilities. The RPN exercise can be continued to drive improvements at each processing step where such impure elements are likely to occur.

3.2. Defect process mapping and yield management system

Defect density is defined as the total number of defects calculated per unit area on the wafer die [4]. In order to reduce defect density between processes, engineers need to identify the specific process steps, equipment, input materials, etc. that are the major contributors to the defect density. This involves the construction of a detailed process flow diagram for isolated segments of the process and the use of various problem-solving tools such as using

http://dx.doi.org/10.5772/intechopen.81058

the Six Sigma concepts, cause and effect diagrams, design of experiments, Pareto principles, etc. to tie the total defects measured at the end of the process sequence to the likely sources in the process flow. Most cutting-edge fabs have automated scripts using machine learning principles to have correlation between in-line defects and final yield loss. Advanced data mining software can quickly scan through very large data sets involving integrated circuit parameters, processing parameters, equipment parameters, probe bins, defect metrics, etc. to come up with various models which the Six Sigma team can use to infer meaningful results during the analyze phase of DMAIC. Thus, having a good defect management and yield monitoring system while benchmarking to industry leaders will enable semiconductor fabs to execute Six Sigma projects efficiently while maintaining a competitive edge in the market.

3.3. Role of SPC in defect monitoring

Another major factor controlling line and probe yields is the ability of the fab to control process variations on critical parameters. In many progressive fabs, situational SPC is now becoming more popular as engineers concentrate more on critical processes and avoid the temptation to overuse tools such as process control charts. Two most important points that fabs must know regarding SPC charts to track in-line defects for Six Sigma scalability are:

i. Due to the high number of processing steps and the possibility of defects forming from any source, fabs must have particle monitoring (PMON) charts which are effectively attribute charts that track defects per million opportunities (DPMO) on bare silicon wafers to plot particles coming from the equipment. This helps to isolate defect sources solely coming from process equipment and can be tied with regular total productive maintenance (TPM) cycles in the fab. Process engineering/metrology experts must be able to decide the sampling frequency in this case.

ii. In-line defect metrology teams must be able to skillfully partition the line in placing SPC charts to control defect metrics based on historical learnings. Most fabs that do not have Six Sigma experts on their team usually have a hard time in determining the number of control charts and end up oversampling or undersampling. Attribute charts for key defect metrics that have downstream product yield impact or customer reliability issues should be given the highest priority in establishing control charts.

4. Case study

This section describes a case study wherein Six Sigma DMAIC methodology was used to tackle a probe yield issue due to an in-line defect contamination occurring in a lithography process step.

4.1. Phase 1: Define

This phase of the DMAIC methodology aims to define the scope and goals of the improved project in terms of customer requirements and to develop a process that delivers these requirements. The first step toward solving any problem in the Six Sigma methodology is by formulating a team of people associated with the process [12]. For the case study in discussion, suspect of in-line process step was not known, and only the critical business impacting factor of yield loss data was known. The initial team comprised of a Certified Six Sigma Black Belt (CSSBB) who were the site quality engineering manager, yield engineering managers, failure analysis engineers, and technicians along with defect metrology engineers. Roles of the team members are shown in the project charter (**Table 1**). Next, the problem statement addressing CTQ and magnitude of the problem was identified. For a span of 13 weeks, one of the factory's key products had been failing for bin fallout along the edges of the wafer resulting in a yield loss of 7% for dies. Failure analysis team was contacted to perform extensive cross-sectional analysis of the defect location. Based on the information available to the Six Sigma team at this stage, problem statement was defined as follows:

Problem Statement: Defects occurring on about 25% of wafers around the edge result in bin failure and die loss on specific technology node.

Project title: Reduction of defects for edge die yield improvement on specific technology node wafers		
Business case for selecting the project:		
Edge die yield loss was ~7% on multiple wafers for a specific technology node. Yield loss on wafers results in microchip failure at the specified region on the wafer which impacts the customers due to poor product reliability.		
Aim of the project:		
Reduce die yield loss to ~1–2% from 7% by reducing in-line defects		
Project champion		CSSBB (quality engineering manager)
Project leader		Head—yield manager
Project team members		Engineering manager—defect metrology Engineer—failure analysis Two process engineers from defect metrology
Characteristics of product/process output and its measure		
CTQ	Measure and specification	Defect definition
Edge die yield	Reduce the number of defective dies at the final probe test by at least >50%	Defective dies at the final probe were cross sectioned and the defect measured 25 μm
Expected customer benefits		High-quality chips within expected time of delivery
Timeline		Define 2 weeks Measure 3–4 weeks Analyze 2 weeks Improve 4 weeks Control 3–4 weeks

Table 1. Project charter.

Here, the CTQ metric of die yield loss could directly be correlated to the number of such defects forming during in-line manufacturing that cause poor yields which results in delayed shipments and dissatisfied customers. The Six Sigma team's next tasks will be outlined in the subsequent sections showing measurement, data analysis, root cause drill down, and corrective action implementation.

4.2. Phase 2: Measure

In this phase, data correlation was conducted to see the correlation between in-line defect counts per wafer to the number of failing dies per wafer at the final probe and percentage of edge die yield loss. Results are shown in **Figure 4**.

It is evident from **Figure 3**, that the correlation between CTQ metric edge die yield losses to the number of in-line defect counts occurring per wafer is linear with the high R^2-adjusted value approaching close to 1. Six Sigma project team must reduce in-line defect count to less than 50 defects per wafer to have no die yield fallout at the final probe test.

4.3. Phase 3: Analyze

The analysis phase consisted of searching through brainstorming rounds, the possible factors that may be affecting the electrical performance of the product. This stage of the Six Sigma process improvement methodology is often termed as Thought Process Mapping [13] wherein process experts and Six Sigma champions assimilate existing facts and data collected so far and look for initial trends and themes to find clues to go after. The factors that were considered most important were raised as hypotheses and tested by several statistical tests.

Wafer fabrication line is partitioned into three modules—front end of line (FEOL), middle of line (MOL), and back end of line (BEOL)—where each module involves complex steps such as lithography, thin-film depositions, etching, planarization, and diffusion. Inspection sampling plans are strategically placed across several processing steps within these three modules considering cost, cycle time, and wafer throughput times. In this case, the project team was

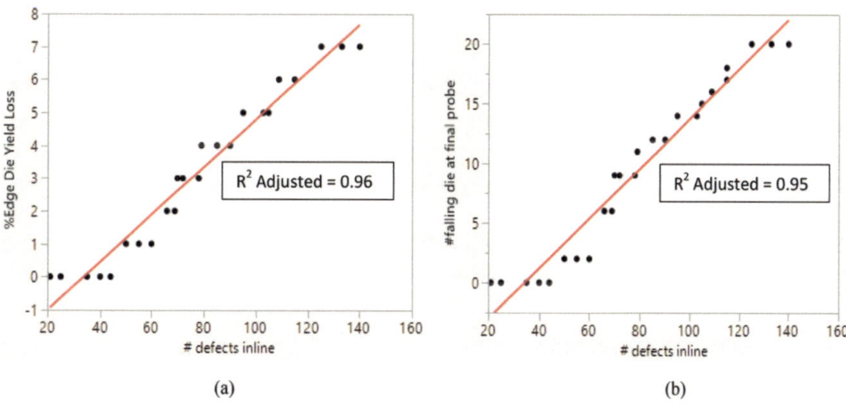

Figure 3. Regression plot of (a) number of failing die per wafer at final probe and (b) percentage edge die yield loss per wafer to number of defects per wafer in-line.

interested to see if there was any in-line defect inspection step that could replicate the defect pattern shown on probe bin wafer map. The project team decided to inspect additional sample wafers through this step and perform failure analysis on the defect locations. More in-line inspection recipes were set up strategically right after metal patterning lithography processes to study defect formation and evolution as the wafers progressed through manufacturing steps.

Through in-line inspections set up across these modules, it was observed that the defect under study was first detected after the metal patterning process. Optical image of the defect was taken along with SEM analysis post BEOL to see defect evolution. This gave an initial indication to the Six Sigma team that metal patterning process and perhaps tool variation in lithography must be analyzed further. There was a need for lithography experts to now help the Six Sigma team in root cause analysis, so lithography process engineering manager, two process engineers, and two process technicians were added to the project team. The project charter was revised to include the new members into the stakeholder team.

Since there is a strong clue of the issue coming from the lithography process area, Six Sigma project team decided to add lithography process experts into the stakeholder team. The project team's next objective was to brainstorm different failure modes that could be occurring with scanner tools in the lithography step and understand root causes from a scientific perspective. For this, the fishbone analysis was used to list several possible root causes across the six Ms—measurement, materials, method, mother nature/environment, manpower, and machine.

Different tools were used to validate the potential causes listed in the fishbone diagram, and a summary of the tools is listed in **Table 2**.

Gemba revealed that there was only one resist supplier to the patterning process, so supplier variation is not a root cause. SEM imaging did not reveal any polymer shearing defects. DOE carried out on resist coating process included a multilevel factorial design where coating speeds were varied from low to high to see if defects could be produced, but it was not the case. Fab contamination studies also showed that particles were well within control and environmental impacts to the formation of in-line defects were negligible. Careful review of all SOPs, manufacturing protocols, preventative maintenance log books, shift pass-downs, etc. did not confirm the validation of any cause. Majority of the effort was then spent in analyzing tool-to-tool variation, which was performed by ANOVA.

For confidentiality purposes, the exact supplier/tool names used in the factory are not mentioned in this chapter. There were three major tools in the factory running this product line, which will be addressed as Tools 1, 2, and 3 for analysis. One-way analysis of variation (ANOVA) of defect metric count with tool set was plotted (see **Figure 4**).

Tool 1 mean value was not only above the target defect count value of 50 but was also significantly above Tools 2 and 3, clearly indicating a problem with this tool.

4.3.1. Tool toggle statistical significance

Since there are multiple tool sets and data is non-normal, the Wilcoxon method is used for multiple tool set comparisons [14]. ($\alpha = 0.05$). H_o: Tool 1 toggle is statistically insignificant; H_a:

Tool 1 toggle is statistically significant. It is observed that p-value is less than 0.05 when Tool 1 is compared with other two tools as per **Figure 5**.

Based on Wilcoxon test, reject H_o and accept H_a. Therefore, Tool 1 toggle is statistically significant and the toggle to defect metric is real. Next, the project team was tasked to look at SPC charts of all critical parameters of all three tools. It was found that the parameter scanner speed and exposure dosage of the immersion hood for Tool 1 were out of control (OOC) and mean value higher than Tools 2 and 3 as per the individuals and moving range (IMR) charts (see **Figure 6**).

Normal quantile plots for all three tools along with process capability index (Cpk) were calculated as shown in **Figure 7**.

SPC charts and normal quantile plots reveal a significant drift in parameter settings for Tool 1 compared to Tools 2 and 3. SPC charts show very tight distribution of points for Tools 2 and 3, but Tool 1 is not only out of spec but also has high amount of wafer-to-wafer variation for scanner speed and exposure dosage.

Out of all the possible failure modes evaluated and tested, it was confirmed that the source of defect could possibly be coming from *Tool 1 metal pattern processing step in lithography area*

Cause	Validation plan
Materials and method related to failure modes	Gemba
	Standard operating procedure (SOP) reviews
	Process engineering change notification reviews
	Process qualification white paper review
	Supplier to supplier variation studies on resist batches supplied
	Perform DOE on resist coating process
	SEM imaging for polymer shearing defect identification
Measurement	Gemba, SOP reviews for specifications, reading confirmations by multiple operators for calibration and solvent volume checks
Environment	Gemba with fab facility department, fab air quality check by reviewing contamination SPC charts, check pressure using fab pressure manometers
Manpower (human errors)	Review preventative maintenance cycle procedures
	Check shift pass-down notes
	Check if any inexperienced operator or new hires joined the team and were performing tasks
Machine	Study tool variation between tools 1, 2, and 3 by conducting ANOVA studies
	Gemba on bubble extraction seal quality and immersion hoods and resist clogging
	Check if any new recipe changes were made before and after the dates when in-line defects were first seen

Table 2. Cause validation plan.

due to significant variation in scanner speed and exposure dosage. The theory behind defect formation due to inaccurate scanner speed and exposure dosage is described below which is a commonly observed phenomenon in the industry [14].

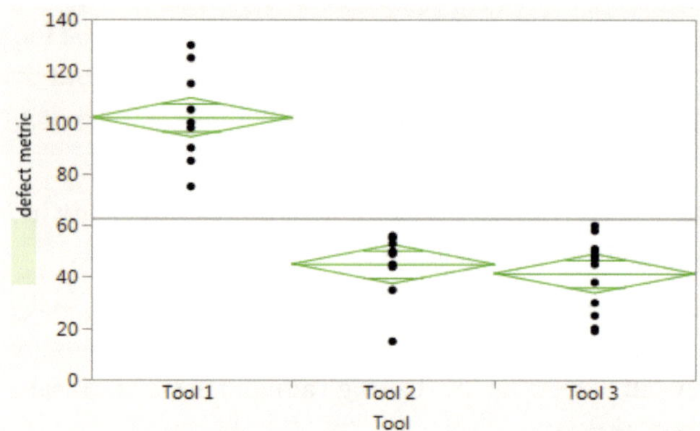

Figure 4. One-way ANOVA for defect metric by tool set.

q*	Alpha
1.95996	0.05

Level	- Level	Score Mean Difference	Std Err Dif	Z	p-Value	Hodges-Lehmann	Lower CL	Upper CL
Tool 3	Tool 2	-0.4615	2.995381	-0.15408	0.8775	-1.0000	-12.0000	8.0000
Tool 2	Tool 1	-12.9231	2.998974	-4.30917	<.0001*	-54.0000	-65.0000	-43.0000
Tool 3	Tool 1	-12.9231	2.998974	-4.30917	<.0001*	-56.0000	-70.0000	-42.0000

Figure 5. Nonparametric test data comparison.

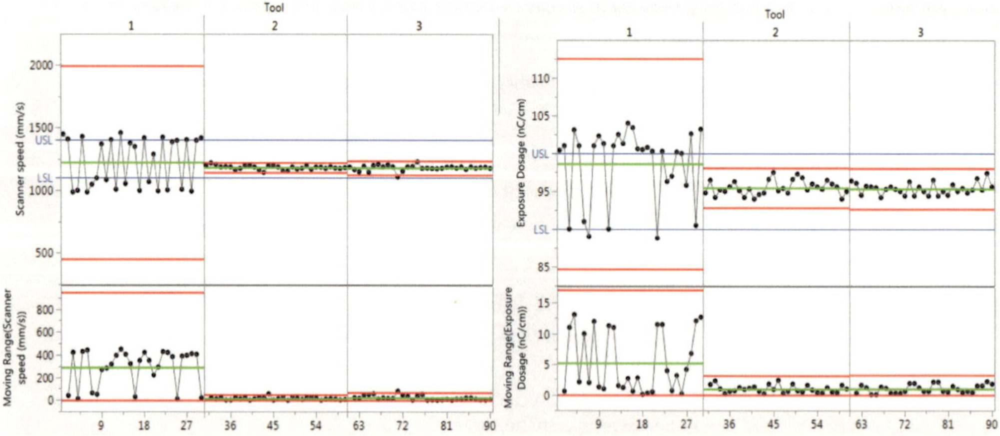

Figure 6. IMR chart comparison for scanner speed and exposure dosage by tool set.

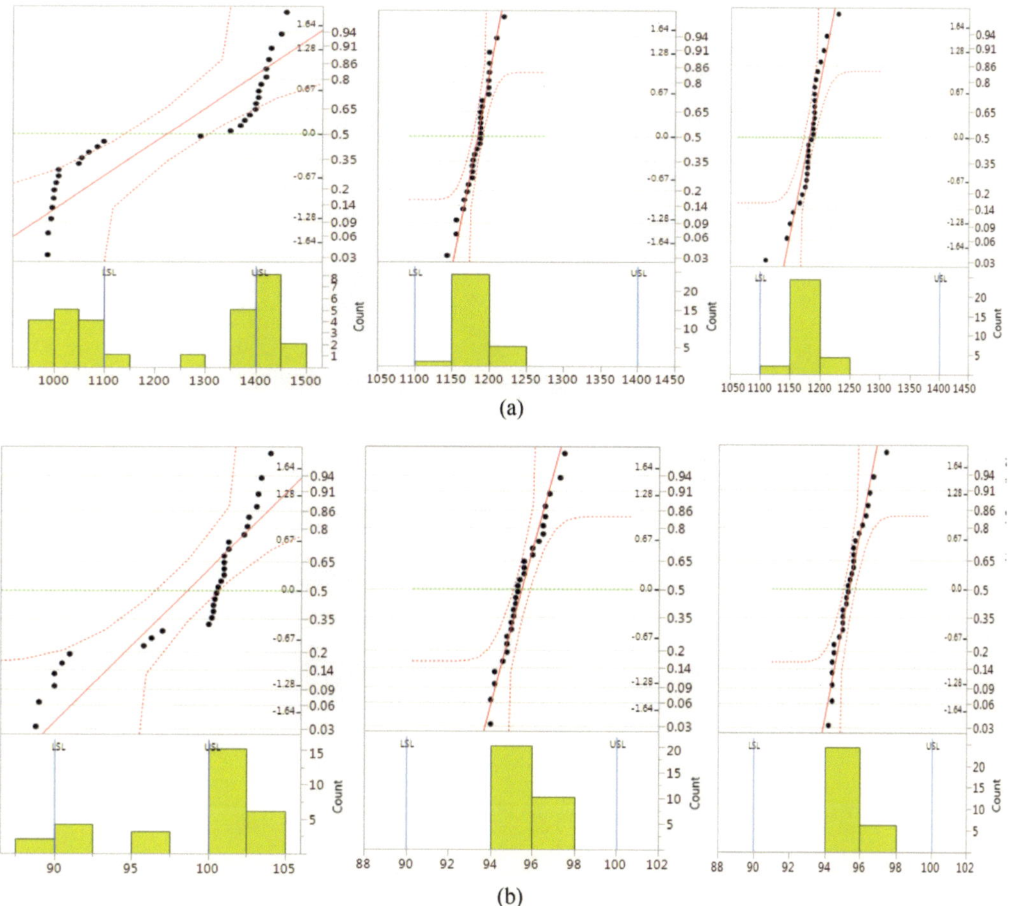

Figure 7. (a) Normal quantile plots for scanner speed and (b) for exposure dosage across three tool sets.

4.3.2. *Defect formation theory*

This defect is caused by a bubble forming in the scanner just prior to exposure. Light passing through the air bubble instead of the immersion water causes light refraction which creates a dipole. Attenuation of light can be seen outside of the ring as shown in **Figure 8** [14].

The system setup usually comprises a wafer placed on the scanner's robotic arm which moves at speeds varying from 1100 to 1400 mm/s per the manufacturer as shown in **Figure 9**. The space between wafer and immersion hoods (where the light source is located) is filled with water as an immersion fluid to increase image resolution by a fraction equal to the refractive index of the fluid [14–16].

The scanner speed of the stage should be optimized to increase productivity without creating defects on the substrate by losing droplets [17]. Immersion liquid level could be a source of a bubble inclusion in the immersion space. Fluid behavior in the region of the recess may cause

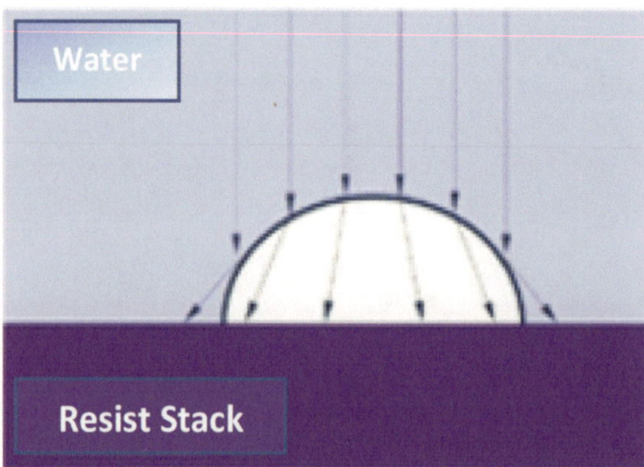

Figure 8. Light diffraction from bubble surface.

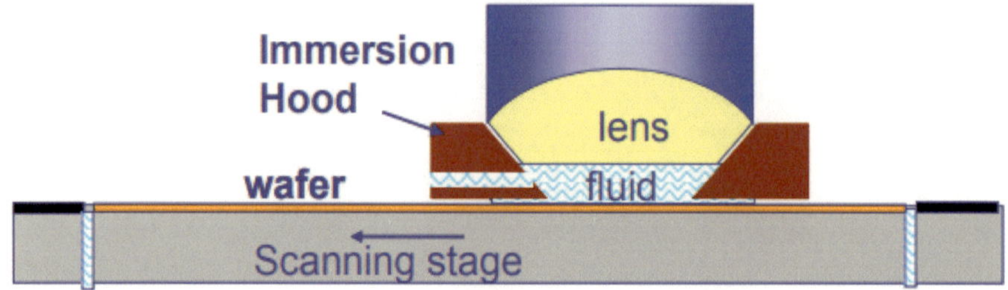

Figure 9. Schematic cross section of an immersion lithographic scanning device [17].

bubbles to form. This bubble may apply a heat load onto a surface onto which it lands, for example, the wafer surface resulting in poor lithographic imaging performance. If the exposure dosage inside the immersion hood is inadequate, bubbles could be entrapped inside the immersion fluid which is water in this case [14–17]. If the scanner moves too fast, the bubble extraction seal (BES) will not have adequate time to suck the bubble out resulting in the bubble being left on the wafer edge. BES extracts the bubbles between the scanner and wafer table.

The Six Sigma project team focused on fixing two main issues here contributing to defect formation—(1) inadequate exposure dosage resulting in air bubble entrapment inside the immersion hood and (2) inadequate scanner speed not giving enough time for BES to suck the bubbles out. Tool 1 settings have drifted significantly from other two tools and has a strong correlation to the high defect counts resulting in yield loss. The next step for the Six Sigma team is to identify optimal tool settings to minimize defect count and have all the three tools operating at these settings. This is discussed in the next phase of Improve.

4.4. Phase 4: Improve

In this phase, optimal tool settings for scanner speed and exposure dosage will be derived using the design of experiments (DOE) full factorial design. Corrective actions addressing the root cause will also be discussed in this section.

Output response is the minimum defect count, while the input factors are scanner speed and tool exposure dosage. Scanner speed values can range from 1000 mm/s to 1400 mm/s per supplier, while the exposure dosage range is from 90 to 100 nC/cm. 16 run DOE table in JMP was created and resulted in prediction profiler obtained after the defect count values for all 16 runs were recorded as shown in **Figure 10**. Using the maximum desirability function in JMP, optimum settings were found to be the scanner speed, 1178 mm/s, and exposure dosage, 94.2 nC/cm.

Based on the above findings for scanner speed and exposure dosage, appropriate interim and long-term corrective actions were proposed and implemented.

4.5. Phase 5: Control

This is the final phase of the Six Sigma DMAIC methodology. Some of the questions that arise after the four phases of problem-solving methodology are as follows: How can one control and monitor the tool parameter settings in line? How can one monitor the in-line defect rate at multiple steps? The simple answer is via statistical process control (SPC) charts. Actual SPC charts on proprietary software have not been shown, but data was exported to JMP has been shown in **Figure 11(a)** and **(b)** post optimal setting discovery and corrective action implementation across all three tools.

It is observed that all three tools have mean values for both the critical parameters very close to the desired settings (1178 mm/s for scanner speed and 94.2 nC/cm for dosage) derived using DOE and have much tighter process control with occasional OOC points which will be addressed real time via the corrective actions implemented. To check process capability post improvements, capability analysis was plotted in JMP (**Figure 12(a)** and **(b)**).

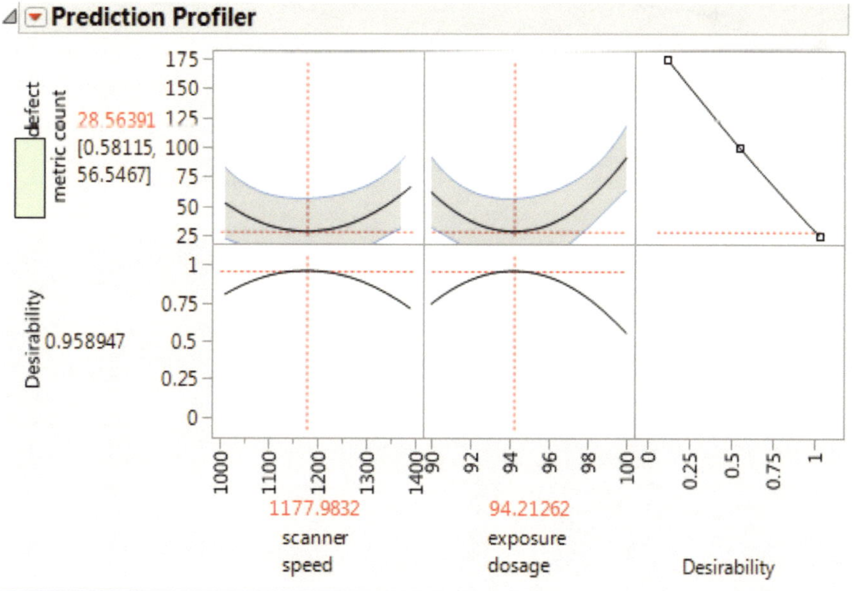

Figure 10. Prediction profiler indicating optimum tool settings for maximum desirability.

Binomial chart plots and Cpk values being >1.33 (1.88 for scanner speed and 2.7 for exposure dosage) clearly indicate high process capability and stability for the lithography metal patterning process post-tool setting optimization.

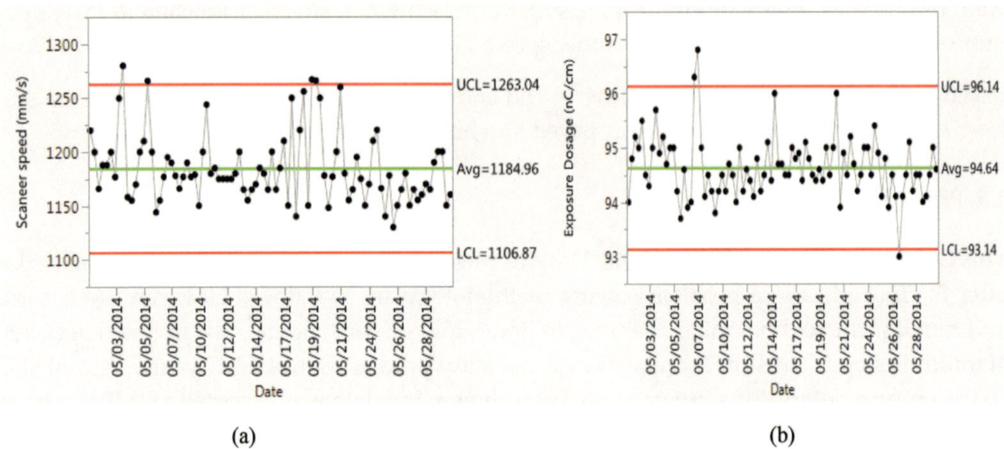

Figure 11. (a) Individual chart for scanner speed and (b) individual chart for exposure dosage.

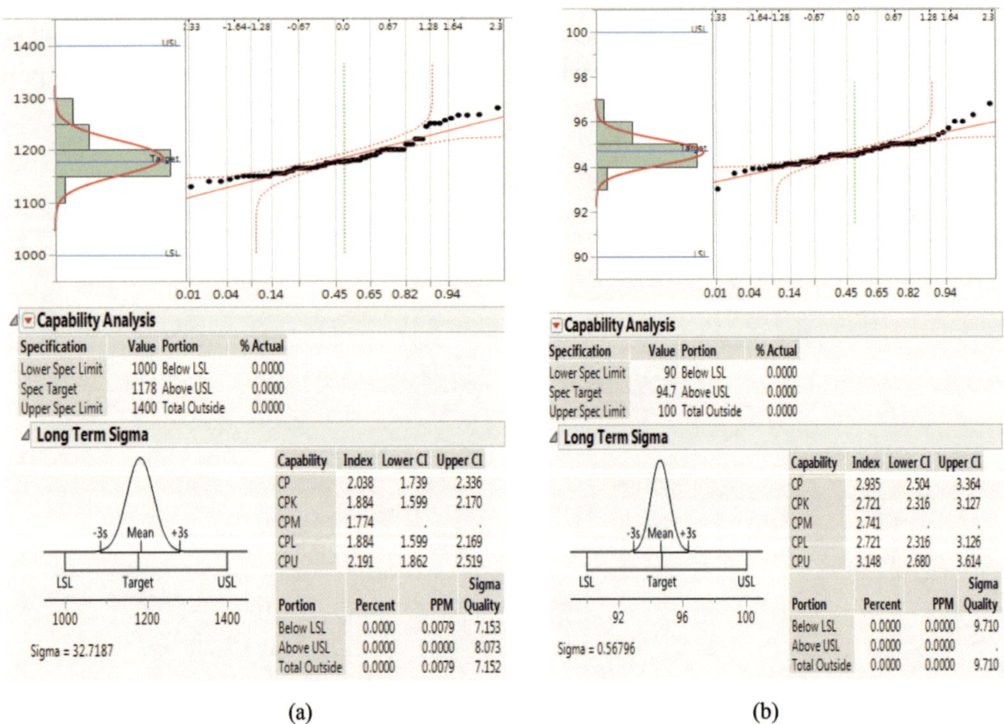

Figure 12. (a) Process capability analysis for scanner speed and (b) process capability analysis for exposure dosage for Tool 1.

FAILURE MODE AND EFFECTS ANALYSIS

Item: Metal Patterning Litho step Responsibility: Quality FMEA number: 123456
Model: XYZ technology node Prepared by: Quality Engineering Page : 1 of 1
Core Team: CSSBB, Lithography Process Engineering, Quality Engineering FMEA Date (Orig): 1/1/2008 Rev: A

Process Function	Potential Failure Mode	Potential Effect(s) of Failure	Sev	Potential Cause(s)/ Mechanism(s) of Failure	Occur	Current Process Controls	Detec	RPN	Recommended Action(s)	Responsibility and Target Completion Date	Action Results: Actions Taken	Sev	Occ	Det	RPN
Wafer Cleaning	insufficient cleaning	left over films from prior steps	7	defects	3	Operator training and instructions	3	63	Review cleaning procedures per SOP, additional training to new technicians	Lithography Shift Supervisors; WW 15-20	Have process technicians undergo certification process every 6 months; add additional inspection steps post cleaning if necessary	5	2	2	20
	high pressure cleaning	wafer breakage	10	scrap	3	Operator training and instructions	3	90	Continuous Improvement	Shift supervisors, engineers	6 month training certification process for technicians	8	2	3	48
	incorrect acid composition in cleaning tank	left over defects, film residues	7	defects	5	None	9	315	Install Composititon detectors; perform regular PM of cleaning tank	Shift supervisors, engineers; WW 15 -24	Installed detectors and establish PM cycle dates	7	3	9	189

Figure 13. PFMEA snapshot for wafer cleaning in patterning process.

4.5.1. *Process FMEA and control plan development*

Six Sigma project team decided to implement FMEA process and has appropriate corrective actions in place as part of control plan. Design-related failure modes and improvements were escalated to the tool manufacturer, while the fab focused on process FMEA for the metal patterning process. **Figure 13** shows the FMEA template for one of the processing steps in the patterning stage (wafer cleaning) which can be used across FEOL, MOL, and BEOL modules. Most of the corrective actions were implemented over 8 weeks, while some of them were part of continuous improvement activities. This FMEA table should be carried out for other sub-process steps involved in the metal patterning process such as resist application, BARC, etc.

Clearly, the new tool settings have greatly reduced the number of in-line defects occurring post-pattern processing steps. The line was then released to run production lots after 8 weeks of improve and control phase actions. This concludes the final step of the DMAIC methodology. The final conclusions and tangible improvement results are discussed in the next section.

5. Conclusions

Post DMAIC, as the mean defect count was continuously controlled and monitored below 50, the final probe yield test data also showed a significant reduction in die yield loss to almost 0% (100% yield) post-DMAIC implementation (**Figure 14**).

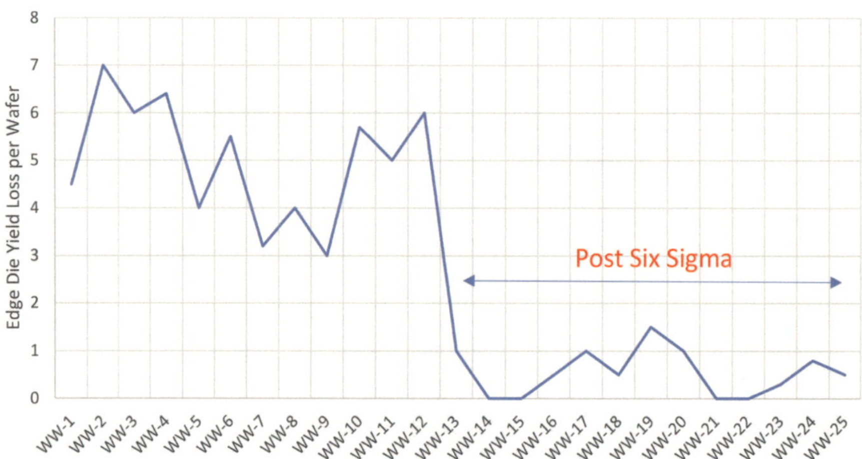

Figure 14. Edge die yield loss graph over 26-week time frame.

It can thus be concluded that DMAIC methodology properly executed under an experienced Six Sigma project team with the support of management is a powerful tool that can reduce process variations and improve product yields, eliminating waste and improving customer satisfaction which ultimately has a significant financial impact on the organization. Six Sigma can be used widely in semiconductor manufacturing environments where there is a tremendous need for defect reduction and tighter process control as the industry advances to smaller technology nodes.

Special note

The author would like to inform the readers that data represented in this paper are not the actual values and closely approximated values have been reported to give an understanding of data analysis, interpretation and application of six sigma methodology.

Author details

Prashant Reddy Gangidi

Address all correspondence to: prg4264@rit.edu

Independent Scientist, California, USA

References

[1] Cavanagh RR, Neuman RP, Pande PS. The Six Sigma Way. New York: McGraw-Hill; 2014

[2] Quality Council of Indiana. CSSBB Primer. West Terre Haute, Ind. Quality Council of Indiana; Vol. 42010. 2014. pp. 125-134

[3] Gutiérrez H, y De la Vara R. Control Estadístico de Calidad y Seis Sigma. México, D.F: Mc Graw Hill Interamericana Editores, S.A. de C.V; 2004

[4] Yield and Yield Management. Integrated Circuit Engineering Corp (ICE). Retrieved from http://smithsonianchips.si.edu/ice/cd/CEICM/SECTION3.pdf

[5] Gijo EV, Scariab J, Antony J. Application of Six Sigma methodology to reduce defects of a grinding process. Quality and Reliability Engineering International. 2011;**27**:1235-1237

[6] ISO/TS 16949:2009. Quality Management Systems. Geneva: ISO; 2008

[7] George ML. Lean Six Sigma. New York: McGraw-Hill; 2002

[8] Breyfogle III. Implementing Six Sigma. 2nd ed. New York: John Wiley & Sons; 2003

[9] Nakajima S. Introduction to TPM: Total Productive Maintenance. Cambridge, MA: Productivity Press; 1998

[10] Wise S, Fair D. The Control Chart Dilemma. ASQ, Knoxville, Tennessee, USA: Quality Progress; 1998

[11] Stapper CH, Rosner RJ. Integrated circuit yield management and yield analysis: Development and implementation. IEEE TSM. 1995;**8**:95-112

[12] Banuelas R, Antony J, Brace M. An application of Six Sigma to reduce waste. Quality and Reliability Engineering International. 2005;**21**:553

[13] DeLayne Stroud J. iSixSigma. 2018. Retrieved from https://www.isixsigma.com/implementation/project-selection-tracking/use-thought-map-increase-efficiency/

[14] Western Electric. Statistical Quality Control Handbook. Indianapolis, IN: AT&T; 1996

[15] Wei Y. Bubble and Antibubble Defects in 193i Lithography. Bellingham, Washington, USA: SPIE; 2008

[16] ASML. US Patent (US7199858). Lithographic Apparatus and Device Manufacturing Method. 2007

[17] Riepen M. Contact Line Dynamics in Immersion Lithography. Netherlands: ASML Research

Chapter 4

Design and Programming of a Wire Winder Device for Extrusion Activity in the Metal-Mechanical Industry

Martha Roselia Contreras Valenzuela,
Alejandro David Guzmán, Diana Lagunas,
Gerardo Vera Dimas, Alina Martínez Oropeza,
Viridiana León-Hernández,
Alber Eduardo Duque Alvarez and
Roy López Sesenes

Additional information is available at the end of the chapter

http://dx.doi.org/10.5772/intechopen.79851

Abstract

The present research is focused on the analysis of an industrial activity where the case under study is wire extrusion activity, which has been developed using old machines. This activity is analyzed considering the task and its elements to identify the movements and sequence of actions performed by the workforce and the machines. One of these activities is the roll-up of lead wire coming from the extruding machine. The task is done by the workforce 73 times during a shift and roughly consists in transporting the wire to the pulley guide and driving the lead wire until the container is rolled up manually. The analysis shows that the workers are exposed to hazardous conditions that could affect their health, meaning they are under risk of suffering an injury or a disease by exposure to repetitive actions and high temperatures (>90°C). Based on the latter, the design and development of a wire winder device has been proposed, which implements a programmable logic controller and servomechanism to replace the activities done manually. A ladder diagram is proposed to control the action performed by the servomechanism based on a stimulus received from the environment.

Keywords: automation, workstation, design, loop closed control, programmable logic controller

1. Introduction

Metal-mechanical industry activities normally include wire extrusion in their processes, which are developed by old machines and involves wasted time and materials as well as costly reworkings [1]. Once the wire is extruded it is normally stored in a coil-shaped container, where it is rolled up ready for the next steps in the value chain based on industrial requirements such as stretched wire. However, the extrusion machines cannot optimize and perform the wire winding activity efficiently, because of the need for human intervention. However, a method to make this task more efficient is possible and can raise production to a satisfactory level. One of the most common solutions consists of the inclusion of automation systems to simplify and improve each part of the process [2], bringing benefits such as reduction in time [3, 4], knowledge, and the ability to control variables such as velocity and temperature, allowing measurement of efficiency in the production line.

Following the trend of using technology to simplify the activities of a manufacturing process [5], the metal-mechanical industry in Mexico began to analyze one of its processes with the aim of improving it, and involved the wire extrusion of a plumb bar as raw material. The machine responsible for performing the wire extrusion activity was old; moreover, most activities are carried out manually by the workforce. An analysis was done to determine the activities involved during wire extrusion, observing that the wire winder task was one of the most critical because the workforce has to roll it manually, which is not ideal because the wire is hot when it leaves the extruder head of the extrusion machine. In addition to this, wire winder activity is performed at a velocity determined by the machine, making it more difficult to handle and satisfy the level of production established by the enterprise. Wire winder activity can be performed and improved with automation technology through a device focused on developing this task, with the capacity to substitute the work carried out manually by the workforce and reduce risk of injuries or diseases [6, 7]. In this way, programmable logic controllers (PLCs) are now being used in the industry around the world with the aim of simplifying and improving the serial instructions present in a sequence of variables that regulate the control system of a device [8], performing a series of actions focused normally on the manufacturing of raw materials in products with high quality. The control programs for PLCs are developed by using relay ladder logic, which is a graphical programming language consisting of software devices (i.e., relays, timing relays, drum sequencers, and programmable counters) to achieve a control strategy [9]. Additionally, the use of 3D design in device automation has become important since it reduces cost and time invested by building a prototype, since is possible simulated and make a prediction behavior of each element based on the freedom grades provide to each servomechanism or articulation [10].

2. Experimental procedure

2.1. Methodology

For the design and programming of the wire winder task of the extrusion process, the following steps were made:

- Perform a task analysis.
- Determine the elements involved in the task carried out manually by the workforce.
- Identify the system variables of activities such as time, temperature, dimension, and workspace.
- Improve the logical sequence of wire winder activity to reduce the wastage of time.
- Conceptually design in 2D a proposed device for wire winder activity specifying the places where the motor elements will be fixed as well as the articulation device for the coil.
- Design in 3D the wire winder, including its mechanical and electrical components.
- Develop the sequence in ladder logic.
- Virtually test the ladder program developed.

2.2. Structural materials

Based on the environment conditions present in the work area, stainless steel 316 was defined as the structural material because of water condensation combined with other chemical compounds such as chlorides, which come from other activities in the facilities, increasing metallic structure degradation. Some of the most important properties of stainless steel 316 are that it contains 16% chromium, 10% nickel, and 2% molybdenum. Molybdenum is added to improve its corrosion resistance against chlorides. Stainless steel 316 is one of the most used materials in industries such as gas, offshore platforms, and many others due to its low price and active passive behavior [11–13]. It was constructed in a rectangular shape to support the electronic and motor components. The design conceptualization of the device includes the use of a coating of paint to be applied to the metallic structure to avoid electrical conductivity in case of cable damage and improve its corrosion resistance [14, 15].

2.3. Hardware

The hardware components used for the automation and design of wire winder activity include the following elements:

1. Three inductive sensors for non-contact detection to ensure the right position of the wire after the extrusion process before it arrives at the feeding cavity in the wire winder device.
2. A PLC based on an Arduino unit was used as a control interface for the logic sequence established for wire winder activity carried out in the extrusion area. Its architecture comprised 20 I/O digitals with a 24 V DC power supply with the capability of interconnecting with a human/machine interface screen system.
3. A relay system was conceptualized to ensure the correct working of the devices and to avoid drops in potential and current.
4. A step servomotor at 180 and 360° was used to execute the changes defined to make the wire winder activity. H bridges where used to control the polarity of the motor at 360°.

2.4. Software

SoapBox Snap Software was used to program the PLC based on an Arduino unit. It included a ladder logic editor and a "soft" runtime straight out of the box. The ladder editor includes standard instructions such as contacts, coils, timers, counters, rising edge and falling edge, and set/reset instructions.

3D drawings were conceptualized and assembled using the Solidworks version 2017 education edition, which is a mechanical program used to create 3D geometry using parametric solids focused on the mechanical design, assembly, and drawings of the workshop.

3. Results and discussions

3.1. Analysis and determination of the elements involved in the wire winder activity of the extrusion process

Extrusion area distribution is shown in **Figure 1**. It is divided into sections: feedstock, extrusion, guide pulley, and storage of wire after rolling (container). The extrusion process is carried out at a pressure of 500 MPa, heating the billet (feedstock) to 575°F as follows:

1. Loading of the billet in the feedstock area of the extrusion machine.
2. Activation of the hydraulic system to start the extrusion of the billet.
3. Manually driving the lead wire extruded to the wire winder area.
4. Winding the lead wire extruded manually by the workforce into the container.

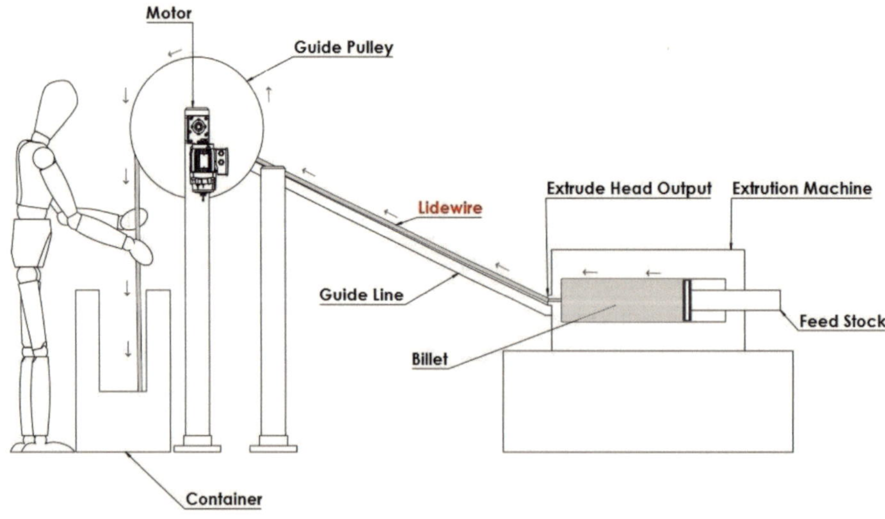

Figure 1. Wire winder element and the devices used to operate it manually.

Analysis of the extrusion process established the elements involved in the wire winder task as follows:

1. Reach.
2. Hold.
3. Move.
4. Roll up.

As was mentioned, the first step to start the extrusion process begins with the loading of the billet, which has a weight of 34 kg inside the feedstock section. Once the hydraulic system is turned on, the extruded product starts to output (lead wire), which is driven manually from the extrusion head output to the guide pulley section. Next, the lead wire begins to roll up manually and is deposited in the storage container by gravity. The devices currently performing the wire winder task have two principal components: a guide pulley and a container for the lead wire after extrusion (**Figure 1**).

The guide pulley has a diameter of 30 cm and is fixed to an AC motor shaft, which is turned on manually before starting to pull the lead wire from the extrusion head output, with the aim of reducing the force applied by the workforce when they pull and guide the lead wire to roll it up. Moreover, the guide pulley is 2 m from the floor making it difficult for the workforce to reach since normally the workforce haas a height average of 1.67 m.

During an 8-h shift, the wire winder task is repeated 73 times with an extrusion cycle time of 9.12 min. To roll up the lead wire, the workforce takes 5.32 min in each extrusion cycle to form circles with a diameter between 60 and 10 cm that go from outside to inside and vice versa in each turn to give stability to the lead wire roll-up.According to the wire winder task analyzed, correct roll-up of the lead wire has a number of important variables:

- Motion sequence for rolling up the lead wire, which is carried out by the worker's arm.
- Time taken for each turn based on its diameter.
- Diameter of each turn, which decreases from outside to inside and increases in the opposite direction.
- Temperature of the lead wire to roll up when coming from the extrude head output.
- Capacity of the container to store the wire roll-up.
- Feed velocity of the guide pulley of the lead wire extruded.

Table 1 shows a comparison of the motion sequence executed by the worker's arm based on the elements identified and the motion sequence that the wire winder device must do to simplify the roll-up task. Furthermore, the time used to perform the roll-up of the lead wire is described with the aim of determining the velocity of the servomechanisms to reach a production level considering the industrial goals.Based on the maximum and minimum diameter defined, an equation is proposed to determine the number of turns that the automatized device must do to finish at least one wire bed as follows:

Task elements	Execution time (min)	Worker's arm	Wire winder device
Reach, move, and hold	5.32	Drive the lead wire to the guide pulley	The lead wire extruded is driven to the guide pulley by an elevator device
Hold		Guide the lead wire to the container	The lead wire is fixed automatically in the arm of the wire winder device to drive it to the container
Hold and move		Move the arm from outside to inside and in the opposite direction forming circles until the lead wire is finished and totally rolled	The device starts to roll up the lead wire when it is detected in the correct position and stops when no more lead wire is detected

Table 1. Comparative analysis of the motions performed by a worker's arm and a wire winder device.

$$n_t = \frac{\varnothing_t}{\varnothing_a} \tag{1}$$

where n_t is the number of total turns, $\varnothing_t$ is the maximum diameter of the roll-up, and $\varnothing_a$ is the wire diameter that corresponds to 0.74 cm. Furthermore, for displacement, the wire from outside to inside reduces the roll-up diameter and the device must start with an angular position of 180° and decrease 2.2° each turn as follows:

$$\theta_m = \frac{\theta_t}{n_t} \tag{2}$$

where θ_m is the decrement in degrees by each turn and θ_t is the initial position in degrees before starting to roll up the lead wire.

The total turns that the device must execute to finish one wire bed are 82 with a decrement of 2.2° by each turn.

The conditional actions that should be followed in a logical sequence so that the device works to the right parameters are described in the flow diagram of **Figure 2**, and are based on the wire roll-up task variables identified.

3.2. Block diagram to establish the task sequence to schedule in a ladder diagram

The block diagram in **Figure 2** describes the principal action needed for the wire winder device to perform in a loop closed system. The reference value is a flush-mountable proximity switch (S1), which detects the presence of the lead wire when it comes from the head extrude area. Once the wire is in the X0 position and a time of 5 s has elapsed, the clamps are closed and the wire transport system is activated carrying the lead wire until the X_1 position, where a second flush-mountable proximity switch (S2) detects the lead wire, turning on the motor M_1 of the guide pulley and sending simultaneously an instruction to the controller to reset M2 and M3 to their initial position (0), ensuring that M2 and M3 are working correctly. When the lead wire is

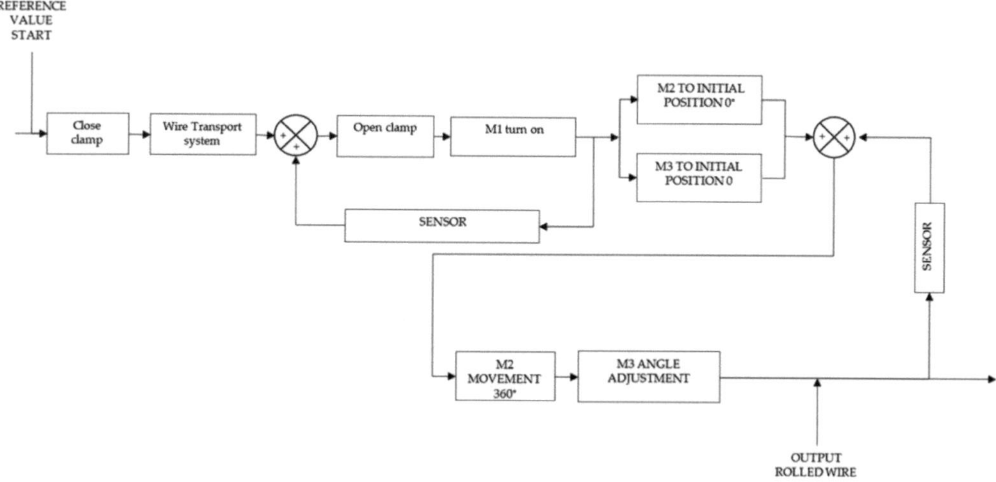

Figure 2. Wire winder element and the devices used to perform it manually.

detected in the X_2 position the flush-mountable proximity switch (S3) closes its contact and in turn M2 starts to spin. As a result M3 increases or decreases its angle to roll up the wire, decreasing or increasing its angle by 2.2° based on its initial position (180 or 0°). The roll-up process ends when the wire is not detected by S2 and S3 in position X1 and X2, respectively, rebooting the system until a new billet has been charged on the feedstock area and new wire starts to be extruded.

As two-push button security system has been configured as a start/stop system in case the workforce has to manually fix the wire winder in positions X0, X1, and X2 or deactivate and reset the system to the initial position when the billet is finished.

3.3. Ladder program in PLC M-Duino 19 R

The electrical components of the wire winder device at a prototype level are described in **Table 2**. A PLC based on an Arduino unit has been selected as controller (**Figure 3**). The PLC M-Duino 19 R is a compact PLC based on open source hardware technology with different input/output units and its principal characteristics are listed in **Table 3**.

The principal symbology used to build the ladder diagram to be loaded in a PLC based on an Arduino unit is described in **Tables 4–6**, declaring the signal inputs that can be defined as an input of electrical stimulation, which can be analog or digital depending of the physical phenomena measured, such as light, sound, pressure, temperature, or the presence of an object. This means that execution of an action by an actuator or software configuration based on the received input allows the logic sequence to produces a response by the controller system through the declared outputs in the ladder diagram. They are in effect electrical floodgates that can open or close the flow of an electrical signal depending of the action to be

Devices	Description
Inputs	
S1	Inductive sensor NPN 24 V DC
S2	
S2	
B1	Push button 22 mm 24 V
B2	
Outputs	
CLAMPS	Servomotor 24 V DC with control driver
MOTOR 1	1/2 HP motor with reducer shaft mounting system 110 V AC
MOTOR 2	1/8 HP motor with reducer shaft mounting system
MOTOR 3	Stepper motor Nema 24 with control driver
SOLENOID 1	Bistable pneumatic solenoid valve 5/2 24 V DC to control a pneumatic rodless cylinder
SOLENOID 2	

Table 2. Electrical components of the wire winder device.

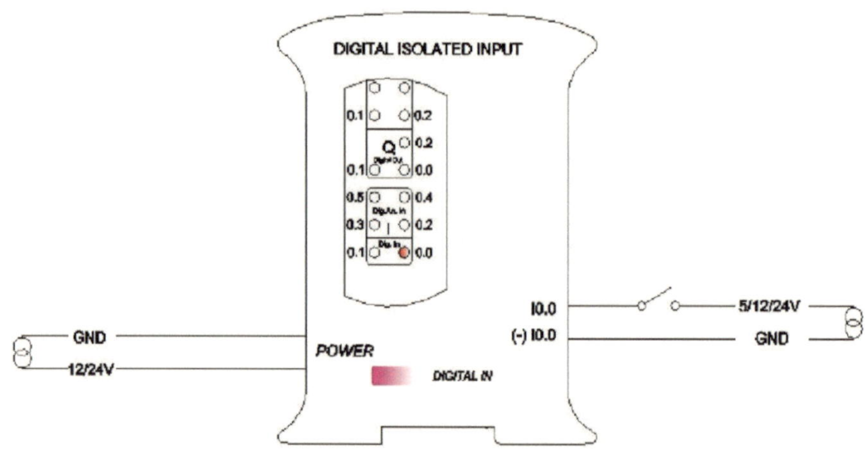

Figure 3. PLC based on an Arduino unit from Industrial Shields.

performed, allowing the servomechanisms configured in a mechanical device to be energized, which in this case correspond to the wire winder. **Figure 4** shows the ladder diagram proposed to control the wire winder device as well as the guide pulley and the wire system transport. Following the condition established in the block diagram (**Figure 2**), the system is turned on when a push button connected to the I.3 input of the relay based on an Arduino unit turns on allowing the auxiliary relay interlock M1 to be energized, which allows timer TT1 to be energized whenever S1 detects the presence of lead wire in the X0 position. Once the I1 contact of the S1 sensor is closed, the timer TT1 is activated, closing its contact T1 after 5 s and

Input voltage	12–24 V DC polarity protection
I_{max}	0.5 A
Size	101 × 119.5 × 70.1 (mm)
Clock speed	16 MHz
Flash memory	256 KB
Communications	I2C1 – Ethernet port – USB – RS485 – RS232 – SPI – (2x) Rx, Tx (Arduino pins)
Total input points	6
Total output points	14
Type of signal	
Analog/digital input 10 bit (0–10 VCC) (5/12/24 V DC)	4
Digital isolated input (24 V DC)	0
Interrupt isolated input HS (24 V DC)	2
Analog output 8 bit (0–10 VCC)	3
Digital isolated output relay	8
PWM isolated output 8 bit (24 V DC)	3

Table 3. Principal characteristics of the PLC M-Duino 19 R.

No.	Symbol	Function	Comment
I1	⊣ ⊢	Digital input	S1
I2			S2
I3			B1
I4			B2
I6			S3

Table 4. Physical inputs.

No.	Symbol	Function	Comment
Q1	–( )–	Digital input	SOLENOID 1
Q2			SOLENOID 2
Q3			CLAMPS
Q4			M1
Q5			M2
Q6			M3

Table 5. Physical outputs.

No.	Symbol	Function	Parameters	Comment
C1	–( )–	Counter	Objective VALUE: 1; CON	N/A
C2		Counter	Objective VALUE: 1; CON	
M1		Auxiliary relay	N/A	E-B1
TT1		Timer	Preset: 5 s; TON	N/A
TT2		Timer	Preset: 10 s; TON	

Table 6. Configurable functions.

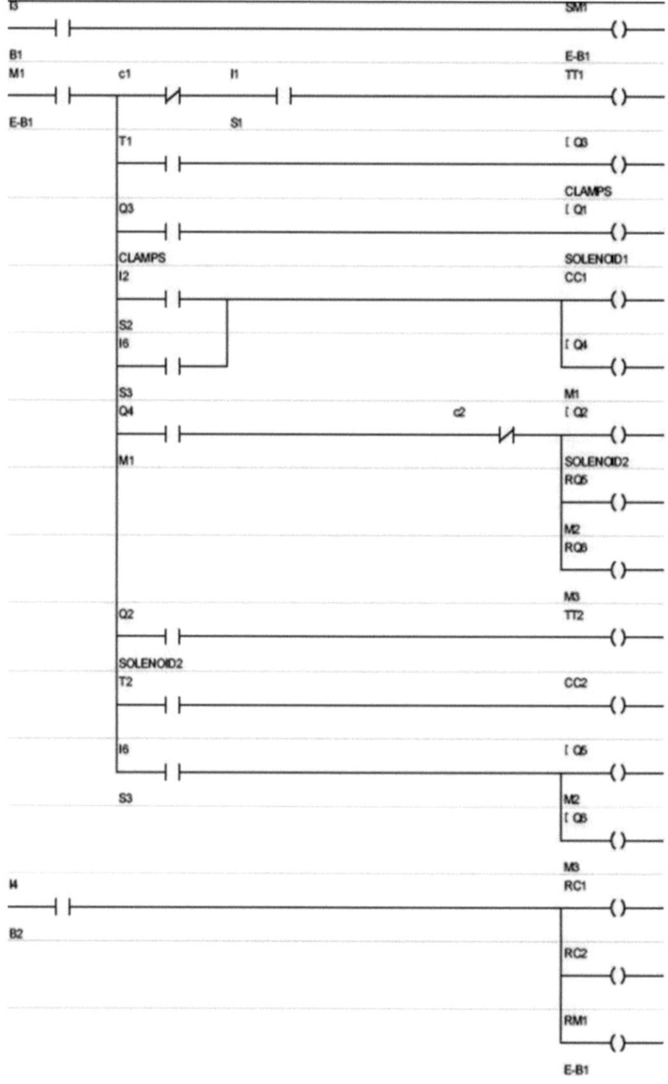

Figure 4. Ladder diagram for the wire winder device.

activating Q3 and Q1 output at the same time, where the first output closes the clamps and the second one turns on the solenoid of a 5/2-way pneumatic valve that activates the wire transport system. When the wire transport system has finished rising in the X1 position the S2 sensor closes its contact activating the I2 input, which closes its contact allowing the 5/2-way pneumatic valve to be turned off, which allows the wire transport system to return to its initial position. At the same time I2 activates the Q4 output closing its contact to turn on the M1 motor of the pulley guide, transporting the wire until the X2 position where the S3 sensor detects the lead wire closes its contact and activates M2 and M3 motors to start the roll-up of the lead wire.

The design proposed for the wire winder device to be implemented in the extrude machine is shown in **Figure 5** where it can be appreciated that the wire transport system, which consists

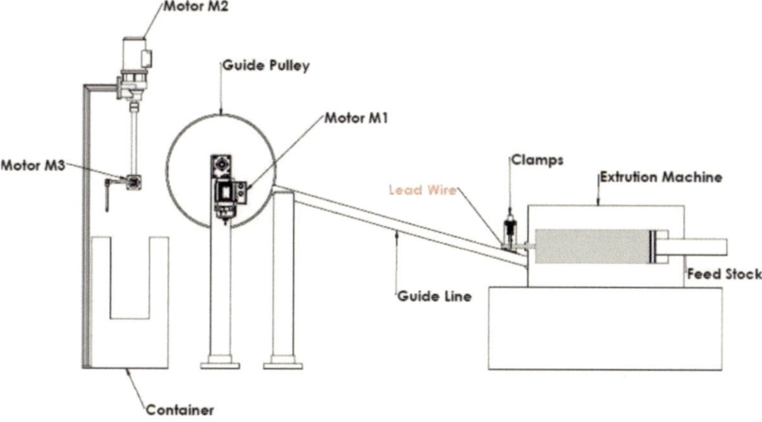

Figure 5. Wire winder device implemented in the extrude machine.

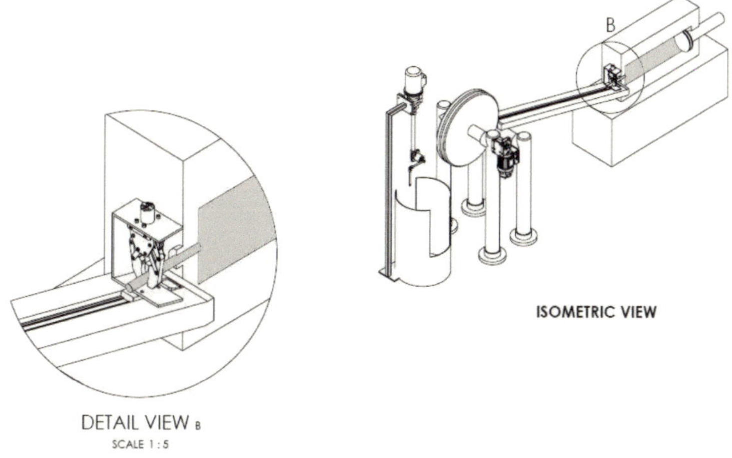

Figure 6. Detail of the clamps fixed over a linear actuator without a stem.

of clamps to take the lead wire coming from the head extrude area (**Figure 6**), after a time is moved to the guide pulley fixed over a linear actuator without a stem, which is activated based on the instruction loaded on the PLC. The wire winder device is operated by two servomotors M2 and M3, which are located after the guide pulley and are activated when the lead wire in the positions previously defined has been detected. Positions X0, X1, and X2 are indicated to understand the working of the new device based on the instruction programmed in the ladder diagram shown in **Figure 4**. It is important to understand that the device is designed to simplify and improve the working conditions in the extrude area to help reduce exposing the workforce to hazardous conditions of high temperatures and repetitive actions.

4. Conclusions

Based on the analysis done it was possible to determine the principal variables of the wire winder system used to conceptualize and design a device with the capacity to perform roll-up of a wire winder after the wire has been extruded. The diameter of each turn was defined, making it possible establish the degrees at which the step motor M2 at 180° should be executed. An equation was determined to understand the angle of reduction in each turn. The split of the wire winder activity was divided based on elements such as reach, hold, move, and roll up. With the aim of avoiding the workforce operating under risky conditions, after extruding a wire transport system was conceptualized mounted over a pneumatic device capable of transporting the wire in a safe and faster way. The sequence of the roll-up task was developed based on previous analysis and allowed the building of a ladder program to control the servomechanism, solenoids, and actuator. A 3D design was proposed to understand the implementation of the wire winder. The device was supported by a stainless steel bar situated to the side of the container two position where declared as balance point to ensure that the roll-up activity will be done.

Author details

Martha Roselia Contreras Valenzuela, Alejandro David Guzmán, Diana Lagunas, Gerardo Vera Dimas, Alina Martínez Oropeza, Viridiana León-Hernández, Alber Eduardo Duque Alvarez and Roy López Sesenes*

*Address all correspondence to: rlopez@uaem.mx

Chemistry and Engineering College, Autonomous University of Morelos State, Cuernavaca, Morelos, Mexico

References

[1] Lela B, Musa A, Zovko O. Model-based controlling of extrusion process. International Journal of Advanced Manufacturing Technology. 2014;**74**:1267-1273

[2] Parasuraman R, Rizzo M. Neuroergonomics: The Brain At Work. Oxford, England: Oxford University Press; 2006. DOI: 10.1093/acprof:oso/9780195177619.001.0001

[3] De Meyer A, Nakane J, Miller Jeffrey G, Ferdows K. Flexibility: The next competitive battle: The manufacturing futures survey. Strategic Management Journal. 1989;**10**:135-144

[4] Endsley MR. Level of automation forms a key aspect of autonomy design. Journal of Cognitive Engineering and Decision Making. 2017;**12**:29-34

[5] Smith MJ, Carayon P. New technology, automation, and work organization: Stress problems and improved technology implementation strategies. International Journal of Human Factors in Manufacturing. 1995;**5**:99-116

[6] Mishev G. Analysis of the Automation and the Human Worker, Connection Between the Levels of Automation and Different Automation Concepts. Sweden: Department of Industrial Engineering and Management, Jönköping School of Engineering. 2006. 71pp

[7] Moniz A. Robots and Humans as Co-workers? The Human-Centred Perspective of Work with Autonomous Systems. Universidade Nova de Lisboa, IET/CICS.NOVA-Interdisciplinary Centre on Social Sciences, Faculty of Science and Technology, Monte de Caparica, Portugal; 2013

[8] Bolton W. Chapter 5—Ladder and functional block programming. In: Programmable Logic Controllers. 5th ed. Boston: Newnes; 2009. pp. 111-146

[9] Zhou MC, Twiss E. Design of industrial automated systems via relay ladder logic programming and Petri nets. IEEE Transactions on Systems, Man, and Cybernetics, Part C (Applications and Reviews). 1998;**28**:137-150

[10] Cusack M. Automation and robotics: The interdependence of design and construction systems. Industrial Robot: The International Journal of Robotics Research and Application. 1994;**21**:10-14

[11] Savaloni H, Agha-Taheri E, Abdi F. On the corrosion resistance of AISI 316L-type stainless steel coated with manganese and annealed with flow of oxygen. Journal of Theoretical and Applied Physics. 2016;**10**:149-156

[12] Singh AK, Chaudhary V, Sharma A. Electrochemical studies of stainless steel corrosion in peroxide solutions. Portugaliae Electrochimica Acta. 2012;**30**:99-109

[13] Xu C, Zhang Y, Cheng G, Zhu W. Corrosion and electrochemical behavior of 316L stainless steel in sulfate-reducing and iron-oxidizing bacteria solutions. Supported by the National Natural Science Foundation of China (No. 20576108). Chinese Journal of Chemical Engineering. 2006;**14**:829-834

[14] Fang X-X, Zhou H-Z, Xue Y-J. Corrosion properties of stainless steel 316L/Ni–Cu–P coatings in warm acidic solution. Transactions of Nonferrous Metals Society of China. 2015;**25**:2594-2600

[15] Assadian M, Jafari H, Ghaffari Shahri SM, Idris MH, Gholampour B. Corrosion resistance of EPD nanohydroxyapatite coated 316L stainless steel. Surface Engineering. 2014;**30**:806-813

Chapter 5

Additive Manufacturing Applied to the Design of Small Satellite Structure for Space Debris Reduction

Jonathan Becedas and Andrés Caparrós

Additional information is available at the end of the chapter

http://dx.doi.org/10.5772/intechopen.78762

Abstract

Space debris has become a major aspect in the last few years. The vast amount of artificial objects orbiting the Earth is increasing. These objects are a threat for active and future missions. Besides, the possibility of uncontrolled re-entry of some of them reaching the surface of the Earth exists. The aim of this work is to provide a view on how to use additive manufacturing technology to design the next generation of satellites in order to reduce the space debris. The components that can be manufactured with additive manufacturing are identified, together with the technologies that are enabled by additive manufacturing to reduce space debris. Finally, the results of these studies and analysis are incorporated into the design of the structure of a small satellite. This study is being part of the H2020 European Project ReDSHIFT (Project ID 687500).

Keywords: additive manufacturing, space debris, satellite structural design, ReDSHIFT H2020 project; CubeSat, impacts mitigation

1. Introduction

Additive manufacturing (AM) is changing the way of designing and manufacturing in multiple sectors. The possibilities that this fabrication method offers compared with the classical ones give it applicability in the space sector. Some advantages of using additive manufacturing technology are the following:

- Possibility of building lots of pieces in short time.
- Weight reduction can be easily achievable with new designs while ensuring structural properties.

- Less environmental impact because of the decrease in time of fabrication and required material. Also, these factors reduce power consumption in fabrication.
- Process speed optimization.
- Part complexity has little impact on manufacturing time and cost plus fewer manufacturing constrains on part design allowing AM to manufacture complex parts. This enables "design for need" instead of "design for manufacturing".
- Creation of composites using printers with double extruder: one for the fiber and one for the matrix. This allows reinforcing selected parts of the components or including specific designs for embedding a bolt or other harness.
- Possibility of embedding wiring or sensors to generate multifunctional structures.
- Applicability to many materials such as metals, composites, polymers or ceramics.
- Some indicate the possibility of in-orbit or on-planet manufacturing [1].

These advantages over classical manufacturing methods indicate that they can be applied in space for new applications or to improve the existing ones. Since the manufacture process with additive manufacturing offers new design possibilities, and also new geometries that are difficult to be obtained with classical methods, it is *a priori* expected that those characteristics can be used to improve the mechanical behavior of a system by establishing new requirements and boundary conditions.

In this work, the use of additive manufacturing is analyzed along with other technologies that this manufacturing method enables. The objective is to implement space debris reduction measures in the design of small spacecraft.

To achieve this objective, additive manufacturing in the space sector is reviewed and analyzed to find which components of spacecraft are susceptible of being manufactured with this methodology. Besides, the technologies enabled by additive manufacturing and applicable to the design of spacecraft are studied. Finally, a design of an 8 U CubeSat structure is proposed, and its impact in space debris reduction is analyzed.

2. Additive manufacturing in space

Considering the advantages of additive manufacturing defined in the introduction, the first objective is to identify which are the applications for additive manufacturing in space. In this section, the main applications of additive manufacturing in satellites are reviewed in order to identify which components of a satellite can be 3D printed and which of them can contribute to space debris reduction.

2.1. Identification of components that can be printed with additive manufacturing

Additive manufacturing has been applied to different components of spacecraft. Some of them are found to have potential to reduce space debris.

2.1.1. Harness

A major challenge in a satellite is the distribution of harness since it is not always easy to place. The design of the spacecraft is based on the payload and tries to minimize the volume of the structure. For that purpose, the components are strategically placed to reduce the internal volume, and harnessing a satellite becomes a major aspect. As a manner to reduce the harness into a satellite, additive manufacturing allows embedding wiring and even sensors into the panels of the structure. At least two options exist to embed electronics into the walls of satellite structures: (1) interrupting the 3D printer in the appropriate layer to place the component or (2) making use of a printer with dual extruder to print a circuit with a conductive material in just one process (see [2, 3]). In these works, the authors contemplate the possibility of embedding an antenna that can be directly printed into the walls of a spacecraft for space-to-ground links. The use of a wall as a backplane for addition of equipment provides multifunctionality to the structure. These applications, although they present some advantages, as for example the use of embedded wiring and sensors can contribute to reduce the debris generated in a catastrophic impact: in principle, the lower the number of components affected by a catastrophic impact, the lower the number of fragments generated. However, they also present major drawbacks such as difficulties to be repaired during the testing or qualification stages of a satellite if needed. This is analyzed in detail in Section 3.2.

2.1.2. Electronics shielding

Protecting sensitive circuitry from the damage caused by the exposure to space radiation is a current problem, which is overcome with the housing of sensitive components inside metal boxes. This, known as shielding, increases the volume and weight of the spacecraft, which are key parameters. The use of additive manufacturing offers an intriguing alternative because the protective metal could be selectively printed to enclose the part, minimizing volume and maximizing protection [4]. Shielding can highly contribute to the reduction of space debris if the satellite is shielded in order to resist impacts of debris particles. Most of these fragments are of size under 1 mm. By designing a shielding capable to resist in orbit the impact of projectiles of this size, the number of generated fragments would be highly reduced and, for instance, the number of space debris particles.

2.1.3. Active thermal management solutions

Additive manufacturing can provide new advances in thermal management; for example, in the fabrication of surface topologies into radiating panels. This solution increases the surface area of the panel and heat pipes embedded into the structure of the spacecraft [3].

2.1.4. Structure

The application of additive manufacturing to the functional structure of the spacecraft can potentially reduce its weight and manufacturing time with respect to classical approaches, such as computer numerical control (CNC) milling. The use of this technology provides more freedom to the designers, who can make use of more complex geometries to improve the structural efficiency without bearing in mind the complexity of manufacturing a part. Small satellites can be benefited with the use of this technology. In The CubeSat Challenge [5], several designs that could be manufactured with additive manufacturing technology were presented. In this case, when designing the structure, the designer shall bear in mind that the structure and all the components of the spacecraft shall be integrated: the reduction of joints can be a major advantage, for example, but it will also make more difficult the integration tasks of the spacecraft components inside the volume of the structure.

However, for large satellites, the use of 3D printing to be part of the functional structure has limitations, mainly because of the reduced 3D printing volume of the metal 3D printers. Then, the use of additive manufacturing for the functional structure of large satellites is nowadays limited to the manufacture of specific components or parts. Otherwise, the advantages of using this technology would be reduced.

2.1.5. Metal brackets

A potential use of additive manufacturing in the functional structure of a spacecraft is the manufacture of metal brackets. The current generation of satellites includes specific brackets used as mechanical interface between the main frame of the satellite and some components such as star trackers, GPS receivers, reflectors and reaction wheels, among others. The main benefits are the weight reduction and the design for need. These advantages have motivated the introduction of additive manufacturing in the manufacture of brackets for satellites, for example the swiss company RUAG optimized the antenna bracket of Sentinel-1A by using this technology [6].

2.1.6. Propulsion

In the space propulsion area, the improvements that come from the application of additive manufacturing are not only focused on mass reduction through new designs of propellant tanks but also on the improvement of the performance. In this case, the possibility of building rocket injectors serves as example of a complex component easily built with a 3D printer with same or even better performances and tolerances than manufactured with classical approaches. One example is the injector for RL-10 upper-stage rocket engine by Aerojet Rocketdyne [7]. This company also used additive manufacturing in the fabrication of a titanium piston, the propellant tank and the pressurant tank of a propulsion system for CubeSat [8].

2.1.7. Space telescopes

NASA's Goddard Space Flight Center analyzed the possibility of assembling a space imaging telescope made almost exclusively from 3D printed components [9].

2.1.8. Shaped antenna reflector

Some antenna reflectors have complex geometries. This makes their manufacture a complex issue. Additive manufacturing can facilitate the manufacture of antenna reflectors for space applications and satellites, making possible the manufacture of the whole antenna in a single piece, independently of the complexity that its geometry can have. The European Space Agency (ESA) developed a 3D printed antenna for satellites in a single piece to be tested in [10]. NASA and Stratasys also developed antennas by using additive manufacturing in [11]. In this last study, in which they validated the technology for space, they claimed that the use of additive manufacturing saved time and money. However, it shall be considered that plastic materials are highly reactive to oxygen atoms present at the operating height of the satellite. This can produce degradation in the component. They solved this problem by painting the components with a high emissivity protective paint to form a glass-like layer on the plastic structure. With this solution, the component can reflect a high percentage of solar radiation and optimize thermal control of the antenna operating conditions.

2.1.9. Fuel tanks

Lockheed Martin used additive manufacturing in the prototyping of fuel tanks, commonly made of titanium, which along with reaction wheels are also a major problem for space debris and which also have a high casualty risk since titanium is very resistant to the effect of the atmosphere in the reentry. They analyzed the design and the manufacture process to develop this component [12]. In that publication, they claimed that additive manufacturing reduced mass, cycle time and material waste.

3. Other technologies for space debris reduction

As described in the previous section, additive manufacturing can be applied itself in several components of a satellite. However, the characteristics of this technology apart from the generally claimed mass reduction, waste reduction, manufacturing time and prototyping, what seems to be really interesting is the (i) design for need, which changes the whole design paradigm previously established in the design for manufacturability and (ii) independence of the complexity of the geometry to manufacture. These two factors facilitate the appearance of new technologies which before additive manufacturing were very difficult to apply, or even impossible, at least in the way they can be applied with additive manufacturing. Two technologies that can highly contribute to the design of satellites to reduce the space debris were identified.

3.1. Lattice/microlattice structures

Additive manufacturing independence of geometry complexity facilitates the manufacturing of metallic lattices, which without this technology were very difficult to manufacture, they being reduced to the creation of foams and other irregular similar structures. However, the capability to generate a lattice with complex although with regular geometry guarantees that the mechanical behavior of such a structure is the same across the whole section. Besides, the lattice can be designed for a specific need to improve or maximize the performance of specific mechanical behaviors. In addition, lattice structures can also reach negative Poisson rates, which increase the resistance of the structure, fracture thoroughness and shear resistance [13].

Boeing created the microlattice variant to lattice, which is the lightest material ever made (microlattice variant is about 99% air) [14].

As reviewed in [15], the most commonly used debris shields for spacecraft rely on several layers with a large standoff distance between them to dissipate the impact energy. However, proven effective, this type of solution is difficult to suit in small satellites in which volume is even more important than mass.

Current solutions are based on structures such as the honeycomb panels. They have poor impact mitigation and may be difficult to integrate in small satellites satisfying, for example, the envelope limits of the CubeSat standard deployment pods. However, lattices can be used to replace the core of the panels and be manufactured integrated with the structure requiring little extra space and at the same time increasing the impact mitigation. For this concept, the energy of the impact is dissipated through plastic deformation and generation of break surfaces; thus, the design can be easily miniaturized. This performance increase was analyzed by NASA through testing [16] by comparing honeycomb panels with open cell foam core sandwich panels (open cell foam has similar mechanical behavior than lattice) concluding that for equivalent panels the impact mitigation was always better for foam core panels.

3.2. Embedded technologies into 3D printed structure

The other set of technologies enabled by additive manufacturing are those linked to embedded wiring and sensors into the 3D printed structure. Some of them can be directly printed or perfectly placed in strategic locations of the structure. Besides, as indicated before, this can reduce the number of fragments if a catastrophic impact occurs in orbit. These technologies were divided into three main groups: embedded devices (such as sensors, electronics and antennas), embedded batteries and embedded wiring. The main limitation of this set of technologies is that it can only be used over not electrically conductive or very well isolated materials. They are reviewed in the following section.

3.2.1. Embedded devices

There are three different approaches to embedded sensors:

- An off-the-shelf device made by traditional methods introduced in the structure during the printing process. In this case, the provided device must be prepared to survive to the printing environment with no damage. In [17], three accelerometers and other electronic devices were embedded in a polymeric matrix with a combination of Fused Deposition Modeling (FDM) and Stereolithography (SLA) additive manufacturing methods.

- An offline 3D printed device introduced in a 3D printed structure. This is similar to the previous case, but instead of embedding an off-the-shelf device, a 3D printed one is embedded instead. An example is the 3D Hall Effect displacement sensor introduced in [18].

- A device printed in the structure. This can be done with the same process or intercalating two or more manufacturing methods. In this case, if a single process is used, the process needs to provide at least two different materials in the same print. Examples of sensors printed in the structure are the 3D printed strain sensors. These consist of injecting a conductive resin in an elastomeric uncured matrix. The final result is a part with an embedded flexible strain sensor [19]. Other application that was included in this classification is that of the antennas that can be printed in a structure or surface: in [20], the authors used Inkjet technology to print a metallic ink (silver based) on convex and concave surfaces. They used conductive meander lines with connected feed lines (printed separately) obtaining performance levels comparable to theoretical results. These 3D printed and miniaturized antennas have multiple applications in addition to classical communication uses.

A review of 3D printing methods in the sensor industry can be found in [21]. It is remarkable that none of the examples described there uses a metallic substrate (structure material) and most of them use printing methods based on polymerization not melting.

3.2.2. *3DP batteries*

Considering the advantage of additive manufacturing referred to the independence of geometry stated above, this technology can be applied to print batteries of any shape. This can be an advantage in satellites because empty spaces in the structure or in the volume of the spacecraft can be used to create a battery that perfectly fits in.

Several Li-Ion battery designs were developed by using additive manufacturing. For example, the authors in [22] used graphene oxide-based ink to print miniaturized batteries, which could be potentially embedded within a spacecraft structure. The capacity of these cells (called 3D-IMA) was 1.2 $mAh{\cdot}cm^{-2}$ normalized with the area of the current collector.

3.2.3. *Embedded wiring*

Embedded wiring for space applications mostly relies on recent conductive inks developed for Inkjet technologies. As stated by Kief et al. in [3], these materials were successfully proven to produce conductive inks for electronics in complex geometries. But low limits in curing temperature led to poor performance in terms of conductivity and carrying capacity which are required for high-power high-frequency applications.

Additionally embedding metallic meshes into polymeric structures were tested, this meshes can act as back planes for electronic components like antennas or as ground planes. Even more meshes can work as support points to weld metallic parts and plastic ones together.

3.2.4. *Analysis of embedded technology*

This analysis shows that although all these 3DP embedded technologies seemed to be promising, they have relevant technical implementation drawbacks. The advantages of using 3D printed embedded technologies, in general terms are associated to the perfect positioning of sensors, elimination of wiring, optimization of space, and reduction of debris fragments but their integrability in a critical system, such as a satellite is still risky: first, any 3D printed conductive element shall be printed on isolated surfaces such as polymers. This obliges satellite manufacturers to come up with new fabrication methods, materials, or additional surface treatment. Second, these technologies present difficulties in any repairing process that can appear during testing or qualification stages. It would be critical that a main sensor or circuit, embedded in a structural element, fails at any stage of the manufacturing or testing processes, obliging manufacturers to create additional parts. The installation of individual sensors that can be repaired or changed with accessibility seems more applicable. Third, the use of multiple extruders should be used in most of cases, with the exception of printing surfaces with conductive inks, as the cases with strain sensors and printed antennas, which could be printed after the metallic part is complete. The printing with different materials is complex and not viable when the thermal properties of the materials being printed substantially differ, as it is the case of many polymers and metals that can be used in space.

However, some of the previous technologies can provide benefits for specific applications:

- 3D printed embedded strain sensors. This solution would provide a perfect positioning of the strain in the surface to be monitored. Furthermore, the surface can be covered, for example with thermal isolation or shielding without affecting the functionality since it is a measure of an intrinsic parameter of the surface in which the strain is placed. Nevertheless, additional analysis and testing should be done in the process of isolation of metallic surfaces and adhesion of the sensor because there are different materials under critical mechanical and thermal loads. This technology would reduce harness and electromechanical components that can be fragmented in case of collision, potentially reducing the space debris of future systems.
- 3D printed embedded antennas. This technology, as described in the analysis, requires an unused area to be printed on. Furthermore, that area cannot be covered with thermal isolation, radiators or shielding. This is difficult to provide in a satellite. However, in some cases, it would be beneficial to manufacture an antenna through an additive manufacturing process that optimizes its shape and performance and then it is installed in the spacecraft as a component afterwards.
- 3D printed batteries. They present the same drawbacks of embedded devices into the satellite structure (if they are embedded in the structure); however, they can be separately printed and integrated in the satellite afterwards. This presents many advantages because they can be printed with any shape, which would be beneficial to make use of any available volume available in the spacecraft or in the structure.

Thus, this analysis indicates that more research and development is needed for 3D printed embedded technologies to reduce the risk of implementation in operative missions.

4. Small satellite structure to reduce space debris

Considering the analysis carried out in the previous sections, an 8 U CubeSat satellite structure was designed to reduce space debris. Notice that there is no mission defined as the design was done to demonstrate the use of different technologies that contribute to space debris reduction, which is the objective of this work. The reasons of selecting an 8 U CubeSat instead of other type were the following:

- To apply additive manufacturing to the whole functional structure of the satellite for demonstration purposes was intended. The 8 U CubeSat is small enough to be fully printed in a typical SLM metal printer such the ConceptLaser M2. This metal printer has a printing volume of 238 × 238 × 230 mm (length × width × height). The 8 U CubeSat has a volume of 200 × 200 × 200 mm, which fits in the 3D printer.
- The 8 U CubeSat follows the CubeSat standard so the analysis carried out in this work can be applied to a large number of satellites (smaller than 1, 1, 1.5, 2, 3 and 6 U), not only to a specific satellite with a specific design.
- The 8 U will facilitate further work on additional analysis of casualty risk of propulsion titanium tanks and reaction wheels.

The technologies implemented in the design of the 8 U CubeSat were additive manufacturing and lattice structures, applied in the structure to improve the shielding. Embedded devices technologies were not considered in this work because of the high risk of implementation in operative mission. Future improvements on those technologies would lead to additional solutions with the benefits already described. Furthermore, the research was focused in the design of the structure, so the implementation of additive manufacturing to other components of the satellite was not addressed, such as metal brackets, harness, propulsion subsystems including propellant tanks, telescopes and antennas. The AlSi10Mg aluminum alloy was selected as the reference material for the structure. The mechanical properties of the material can be funded in [23].

This work does not enter into details of the structural design of the 8 U CubeSat by following the CubeSat standard. This can be found in [24]. It is focused on the design of the lattice core panels.

4.1. Lattice panel concept

Following the work carried out by NASA in [25], the geometry of the lattice core panel was defined to provide the best impact efficiency possible. The panel was designed as a sandwich panel concept. It was divided into three parts: (1) an inner panel, (2) lattice core and (3) shear panel. Because of the benefits of using additive manufacturing, the inner panel and the lattice core can be printed together. The shear panel can be assembled afterwards with bolted unions. The three parts were not printed together because the SLM printer used metal powder. If the volume to be printed was closed, the residual powder could not be extracted from the part, remaining inside and changing the mechanical properties of the part. **Figure 1** shows the concept of the panel with lattice core. For instance, the structure was constituted by six independent faces with lattice core panel.

A common CubeSat usually has the maximum thickness of 7.7 mm and with 1 mm to assemble shear panels. Thus, the total margin to increase the shielding and width of the structure was 8.7 mm. Increasing this margin would limit the incorporation of standard COTS components in the satellite and would also difficult the integration of the spacecraft in a POD, which is the common interface with the launcher for CubeSats. So excluding the shear panel, the lattice thickness plus the inner wall added could not be wider than 7.7 mm.

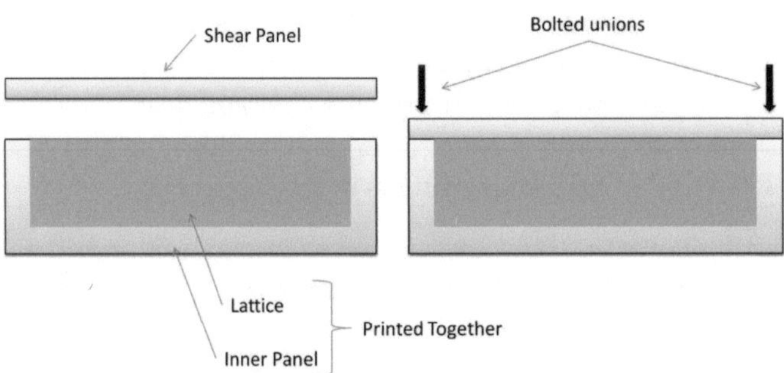

Figure 1. Lattice core panel concept.

4.2. Lattice core panel design

To design the panel, a three-dimensional cost function is generated by relating (1) the impact area efficiency when a projectile of aluminum alloy impacts the surface with an angle of 0° (i.e., perpendicular to the impact surface and a velocity of 8 km/s magnitude) with (2) lattice relative density and with (3) lattice thickness. Eq. (1) was adapted from [25]:

$$d_c = 1,915\frac{(t_w + 0,5AD_{latt}/\rho_w)^{2/3}{t_{latt}}^{0,45}(\sigma/70)^{1/3}}{{\rho_p}^{1/3}{\rho_f}^{1/9}(V)^{2/3}\cos(\theta)^{4/5}} \tag{1}$$

where d_c is the critical projectile diameter at shield failure in cm, t_w is the rear wall thickness in cm, AD_{latt} is the area density of the lattice core in g/cm^2, ρ_w is the density of the rear wall in g/cm^3, t_{latt} is the thickness of the lattice in cm, σ is the rear wall at 0.2% offset tensile yield stress in ksi (kilopound per square inch), ρ_p is the density of the projectile in g/cm^3, ρ_f is the density of the shear panel in g/cm^3, V is the impact velocity in km/s and θ is the impact angle from the target normal vector.

$$AD_{latt} = \rho_{LRel}\rho_{AlSi10Mg}t_{latt} \tag{2}$$

where $\rho_{AlSi10Mg}$ is the AlSi10Mg aluminum alloy density in g/cm^3, and ρ_{LRel} is the relative density of the lattice from 0 to 1.

$$\rho_A = \left(\rho_{AlSi10Mg}(t_w + t_s) + AD_{latt}\right) \tag{3}$$

ρ_A stands for the area density of the lattice core panel in g/cm^2 and t_s is the shear panel thickness in centimetre.

$$\mu_I = \frac{d_c}{\rho_A} \tag{4}$$

where μ_I is the impact area efficiency in cm^3/g.

Figure 2 depicts the results of the optimization region. From this region it was concluded that the optimal solution was a lattice of 10% relative density. Ideally, a higher reduction of the

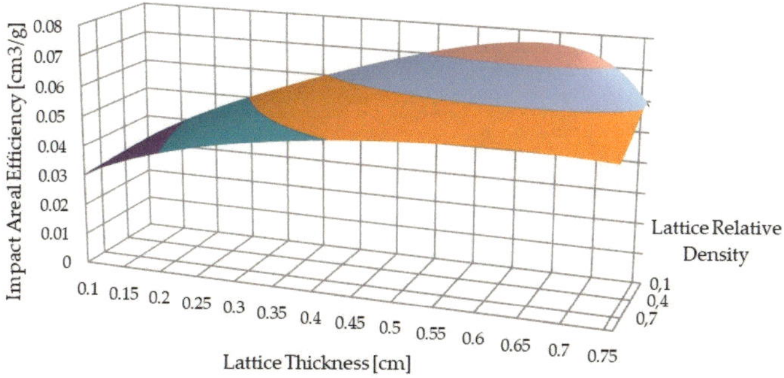

Figure 2. Lattice core sandwich impact areal efficiency [cm^3/g].

lattice relative density would increase the efficiency, but there is a physical limit as the printer has limited resolution and the model used to calculate the critical projectile diameter is not valid since the mechanic behavior changes for lower relative densities.

Figures 3 and **4** depict the lattice thickness in function of the impact aerial efficiency defined in Eq. (4) and the critical projectile diameter defined in Eq. (1) with a lattice relative density of 10%. The design point was chosen to have a lattice thickness of 0.35 mm, a relative lattice density of 10%, and an inner panel with 0.42 mm thickness. This point is not the maximum of the curve but by having moved the design point slightly to the left, although there is lower impact aerial efficiency, there is higher resistance to impacts of projectiles of larger size. The decision of maximizing the impact resistance instead of the efficiency was taken due to the fact that the majority of the space debris has a size lower than 1 mm. In addition, these fragments cannot be tracked so impact avoidance maneuvers cannot be done [26, 27]. Consequently, the Impact Areal Efficiency for the design point was $5.97 \times 10^{-2} cm^3/g$, the Critical Projectile Diameter was 8.9×10^{-2} cm and the designed panel mass is 0.60 kg.

4.3. Lattice core panel analysis

In this section, the shielding performance of the lattice core panel designed is compared with three cases:

- A classical shear panel of 1 mm thickness. This solution weights 0.11 kg.
- An IsoMass panel: it keeps the same mass than the lattice core panel. It has larger thickness than the 1 mm shear panel and generates an equivalent solid plate to that of the lattice core panel. This solution measures the benefit of making a more complex geometry which cannot be made with traditional methods. For a 20 × 20 cm plate, this solution weights 0.60 kg.
- IsoVolume panel: CubeSats are restricted in volume to take advantage of COTS pods so a second variant is presented, instead of keeping the same mass now a solid plate with the same volume as the panel with lattice is defined. This approach evaluates how volume efficient the lattice core solution is compared to a heavier solution with its same volume. For a 20 × 20 cm plate, this solution weights 0.94 kg.

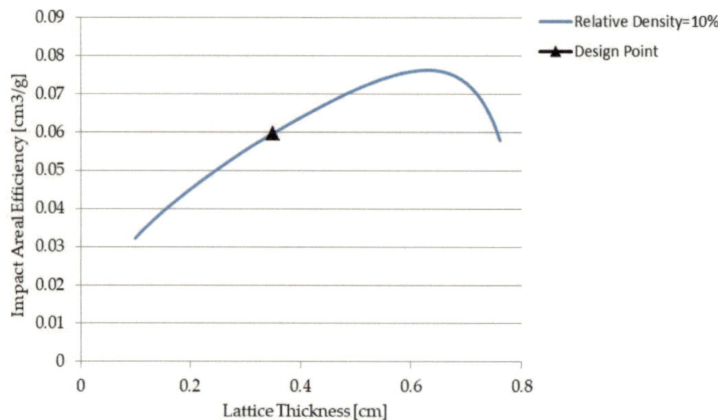

Figure 3. Impact areal efficiency vs. lattice thickness for a relative density of 10%.

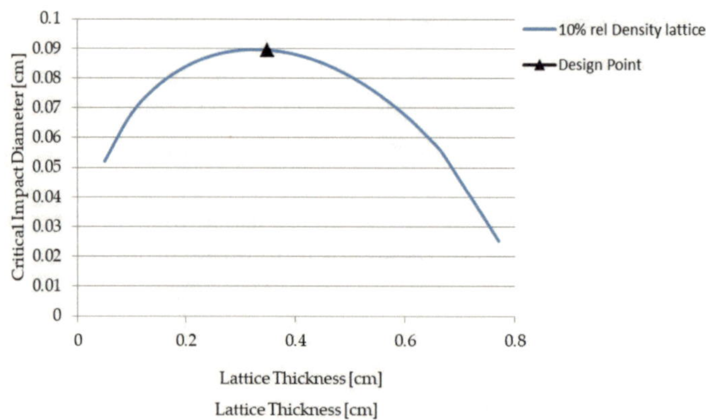

Figure 4. Critical projectile diameter vs. lattice thickness for a relative density of 10%.

The different plates and the lattice core critical projectile diameters, d_c, for a span of impact velocities can be obtained by implementing the equations introduced in [16] for the single plates and in [25] for the lattice core panel. The relative impact mitigation (RIM) of the plates compared with the lattice can be obtained with the following equation:

$$RIM = \frac{d_{c_{plate}} - d_{c_{lattice}}}{d_{c_{lattice}}} \times 100 \tag{5}$$

where $d_{c_{plate}}$ is the critical projectile diameter at shield failure for the solid plate and $d_{c_{lattice}}$ the same parameter for the lattice core panel.

Figure 5 shows the critical projectile diameter at shield failure for all the four panels. For velocities of the projectile lower than 4.6 km/s both the IsoMass and the IsoVolume panels resist to projectiles with higher diameters than the lattice core panel. At this speed both the IsoMass and the lattice core panels can resist impacts of projectiles with 0.13 cm size, while the IsoVolume panel can resist impacts with projectiles of 0.20 cm size. However, from 4.5 km/s, the lattice core panel resists to larger size impacts than the IsoMass panel, and from velocities

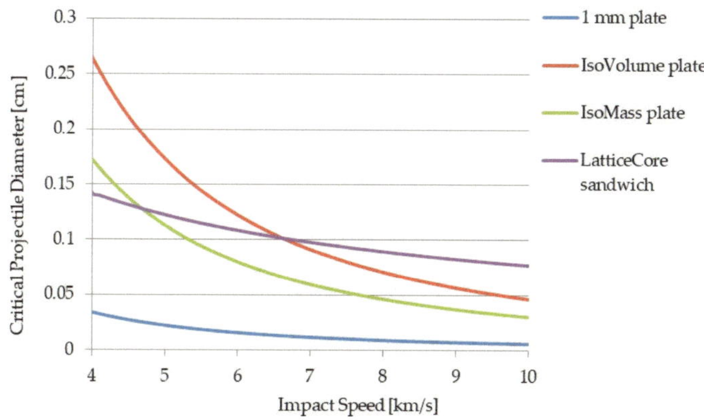

Figure 5. Plates vs. lattice core critical projectile.

higher than 6.4 km/s (at which both lattice and IsoVolume panels resist impacts of projectiles with 0.1 cm size), the lattice panel has more resistance than the IsoVolume panel to impacts of projectiles with larger size. This result is notorious since the IsoVolume panel having more mass than lattice has lower resistance to impacts. In addition, the classical panel with 1 mm width has lower resistance than the lattice core panels in all conditions, as could be expected.

Figure 6 shows the relative impact mitigation in percentage of the plates with regard to the lattice core panel. Even though an optimized lattice impact mitigation of only 0.089 cm at 8 km/s may seem too low, when compared to solid plates results are remarkable. A simple shear panel shields an 80% less than the lattice core panel for the hypersonic regime while the mass is only a 55% higher; compared then with the solution with the same mass the lattice core panel performs better for high speed impacts which are the impacts potentially more dangerous: 7 km/h or higher. The lattice core for these impacts outperforms the IsoVolume plate, which is an 80% heavier solution.

Finally, the designed 8 U CubeSat structure can be seen in **Figure 7**.

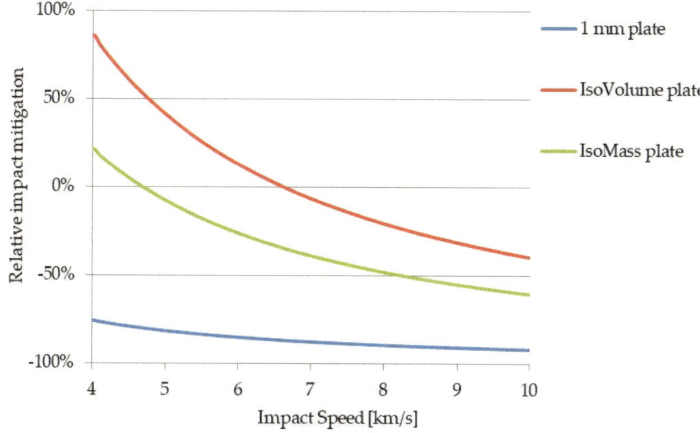

Figure 6. Relative impact mitigation.

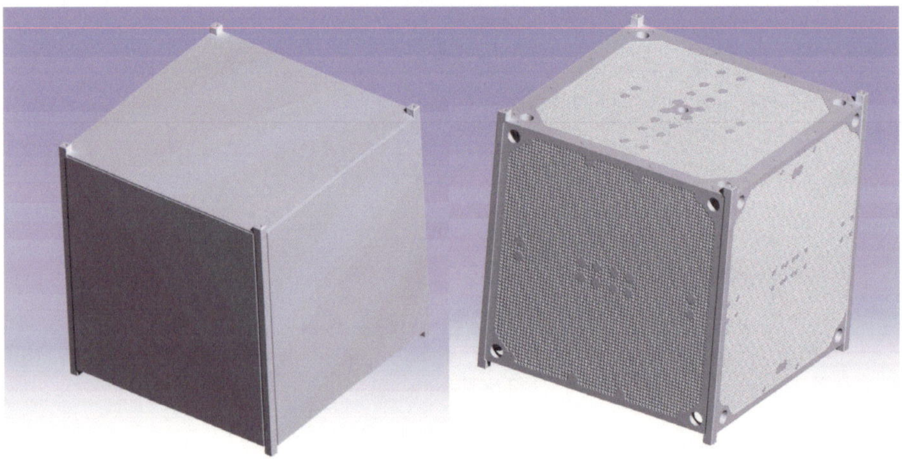

Figure 7. Isometric view of the small satellite structure with and without shear panels (left and right views).

5. Conclusions

In this chapter, a review of the additive manufacturing technology applied to satellites was done. Besides, the applicability of other technologies that can be enabled by this manufacturing method was also analyzed. As a consequence of the study, it was found that the application of additive manufacturing and lattice structures could be applied to improve the behavior of a satellite to reduce space debris when these technologies were incorporated in the functional structure of small satellites and in the impacts shielding of the system. Then the structure of an 8 U CubeSat was proposed and designed incorporating a sandwich panel with lattice core. The design was analyzed and compared with classical CubeSat panels of 1 mm thickness, with an IsoMass panel (i.e., same mass than the lattice core panel) and with an IsoVolume panel (i.e., an aluminum panel with the same volume than the lattice core panel but with 56.7% more mass). It was found that the lattice core panel in impacts with particles at velocities higher than 4.6 km/s provides more shielding than the IsoMass panel and in impacts with higher velocity than 6.4 km/s provides more shielding than with the IsoVolume panel.

For instance, the improvement in the impact shielding of a spacecraft can dramatically reduce the space debris by designing the future satellites accordingly. If they resist to a larger number of impacts, new fragments of space debris will not be generated. According to National Research Council [26], the highest population of space debris within 1600 km of the Earth surface is constituted of small size fragments lower than 1 mm diameter. The authors estimate that hundreds of trillions of fragments under this size are orbiting and impact at velocities with magnitudes between 6 and 8 km/s. They can be potentially destructive since objects of this size cannot be tracked. On the other hand, they estimate that approximately the order of magnitude of larger fragments is 10 millions. However, objects with size between 1 and 5 cm and higher can be tracked, so collision avoidance maneuvers could be done to avoid impacts [27]. This indicates that the proposed design can resist impacts of hundreds of trillions of debris fragments, in the order of magnitude of 10,000 fragments can be tracked so collision avoidance manoeuvers could be executed (most destructive ones) and that the design would be exposed to approximately 1–10 million fragments of size between 1 mm and 1 cm sizes.

Acknowledgements

The research leading to these results has received funding from the Horizon 2020 Program of the European Union's Framework Programme for Research and Innovation (H2020-PROTEC-2015) under REA grant agreement number 687500 – ReDSHIFT (http://redshift-h2020.eu/).

The authors want to acknowledge Irene Vázquez and Gerardo González for their support to identify the applications of additive manufacturing in space.

Conflict of interest

This publication reflects only the author's views and the European Commission is not liable for any use that may be made of the information contained therein.

Author details

Jonathan Becedas*† and Andrés Caparrós†

*Address all correspondence to: jonathan.becedas@elecnor-deimos.com

Elecnor Deimos Satellite Systems, Puertollano, Spain

†The two authors equally contributed to the paper

References

[1] Ghidini T. An overview of current AM activities at the European Space Agency. In: 3D Printing & Additive Manufacturing—Industrial Applications Global Summit 2013, London, UK; 2013

[2] Shemelya C, et al. Multi-functional 3D printed and embedded sensors for satellite qualification structures. Sensors, 2015 IEEE, Busan, South Korea, 2015. pp. 1-4

[3] Kief CJ et al. Printing Multi-Functionality: Additive Manufacturing for CubeSats. In: Proceedings of the AIAA Space 2014 Conference and Exposition; San Diego, USA; 2014

[4] Keesey L. Goddard's Storied Tradition Potentially Expanded Through 3-D Manufacturing [Internet]. 5 February 2014. NASA's Goddard Space Flight Center, Greenbelt, Maryland, USA. Available from: https://www.nasa.gov/content/goddard/storied-tradition-potentially-expanded-through-3-d-manufacturing/#.WAXxu-B96Uk (Accessed: 21 March 2018)

[5] The Cubesat Challenge [Internet]. 2015. Available from: https://grabcad.com/challenges/the-cubesat-challenge (Accessed: 21 March 2018)

[6] Aerospace: RUAG–Additive Manufacturing of Satellite Components [Internet]. Available from: https://www.eos.info/case_studies/additive-manufacturing-of-antenna-bracket-for-satellite (Accessed: 21 March 2018)

[7] Aerojet Rocketdyne successfully tests 3D printed injector for RL-10 upper-stage rocket engine [Internet]. 9 March 2016. Available from: http://www.3ders.org/articles/20160309-aerojet-rocketdyne-successfully-tests-3d-printed-injector-for-rl-10-upper-stage-rocket-engine.html (Accessed: 21 March 2018)

[8] Aerojet Rocketdyne Successfully Demonstrates 3D Printed Rocket Propulsion System for Satellites [Internet]. 15 December, 2014. Available from: http://www.rocket.com/article/aerojet-rocketdyne-successfully-demonstrates-3d-printed-rocket-propulsion--system-satellites (Accessed: 21 March 2018)

[9] Keesey L. NASA Engineer Set to Complete First 3-D-Printed Space Cameras [Internet]. NASA's Goddard Space Flight Center, Greenbelt, Maryland, USA. 6 August, 2014. Available online: https://www.nasa.gov/content/goddard/nasa-engineer-set-to-complete-first-3-d-printed-space-cameras/#.WAX9fOB96Ul (Accessed: 21 March 2018)

[10] Szondy D. ESA puts 3D-printed satellite antenna to the test [Internet]. 20 March, 2016. Available from: https://newatlas.com/esa-3d-printed-satellite-antenna/42373/

[11] Stratasys Direct Manufacturing Builds the First 3D Printed Parts to Function on the Exterior of a Satellite [Internet]. Available from: https://www.stratasysdirect.com/resources/case-studies/3d-printed-satellite-exterior-nasa-jet-propulsion-laboratory (Accessed: 21 March 2018)

[12] Crawford M. Lockheed Launches 3D Printing into Space, ASME.org, August, 2014. Available online: https://www.asme.org/engineering-topics/articles/manufacturing-processing/-lockheed-launches-3d-printing-into-space (accessed on 21st March 2018)

[13] Gong X, Anderson T, Chou K. Review on powder-based electron beam additive manufacturing technology. Manufacturing Review. 2014;**1**:2. DOI: 10.1051/mfreview/2014001

[14] The Lightest Metal Ever. Available online: http://www.boeing.com/features/2015/10/innovation-lightest-metal-10-15.page (Accessed: 20 March 2018)

[15] Christiansen EL. Meteoroid/debris shielding. In: Lyndon B, editor. JSC Technical Report. Houston, Texas, USA: National Aeronautics and Space Administration, Johnson Space Center; 2003

[16] Ryan S, Christiansen EL. Micrometeoroid and Orbital Debris (MMOD) Shield Ballistic Limit Analysis Program, NASA Johnson Space Center, NASA/TM–2009–214789, 2009a, February, 2010

[17] Macdonald E et al. 3D printing for the rapid prototyping of structural electronics. IEEE Access. 2014;**2**:234-242. DOI: 10.1109/ACCESS.2014.2311810

[18] Bodnicki M, Pakuła P, Zowade M. Miniature displacement sensor. In: Advanced Mechatronics Solutions. Springer International Publishing; 2016. pp. 313-318

[19] Muth JT et al. 3D printing: Embedded 3D printing of strain sensors within highly stretchable elastomers. Advanced Materials. 2014;**26**:6307-6312

[20] Adams JJ et al. Conformal rinting of electrically small antennas on three-dimensional surfaces. Advanced Materials. 2011;**23**:1335-1340

[21] Xu Y et al. The boom in 3D-printed sensor technology. Sensors. 2017;**17**(5):1166. DOI: 10.3390/s17051166

[22] Fu K et al. Graphene oxide-based electrode inks for 3D-printed lithium-ion batteries. Advanced Materials. 2016;**28**(13):2587-2594

[23] AlSi10Mg ConceptLaser Datasheet. Available online: https://www.concept-laser.de/fileadmin//user_upload/Datasheet_CL_31AL.pdf (Accessed: 10 May 2018)

[24] 6U CubeSat Design Specification, The CubeSat program, Cal Poly SLO, California, USA, 2016. Available online: https://static1.squarespace.com/static/5418c831e4b0fa4ecac1bacd/t/573fa2fee321400346075f01/1463 788288448/6U_CDS_2016-05-19_Provisional.pdf (Accessed : 23 March 2018)

[25] Yasensky J, Christiansen EL. Hypervelocity Impact Evaluation of Metal Foam Core Sandwich Structures. Houston, Texas, USA: NASA Lyndon B. Johnson Space Center; 2007

[26] National Research Council. Orbital Debris: A Technical Assessment. Chapter 3. Washington, DC: The National Academies Press; 1995. DOI: 10.17226/4765

[27] Space Debris Basics. Available online: http://www.aerospace.org/cords/all-about-debris-and-reentry/space-debris-basics/ (accessed on 23rd March 2018)

Chapter 6

Multiple Criteria Evaluation of Assembling Buildings from Steel Frame Structures

Ruta Miniotaite

Additional information is available at the end of the chapter

http://dx.doi.org/10.5772/intechopen.79858

Abstract

Steel frame structures are often used in the construction of public and industrial buildings. They are used for all types of slope roofs; walls of newly built public and industrial buildings; load bearing structures; and roofs of renovated buildings. The process of assembling buildings from steel frame structures should be analysed as an integrated process influenced by factors such as construction materials and machinery used, the qualification level of construction workers, complexity of work, and available finance. It is necessary to find a rational technological design solution for assembling buildings from steel frame structures by conducting a multiple criteria analysis. The analysis provides a possibility to evaluate the engineering considerations and find unequivocal solutions. The rational alternative of a complex process of assembling buildings from steel frame structures was found through multiple criteria analysis and multiple criteria evaluation. In multiple criteria, evaluation of technological solutions for assembling buildings from steel frame structures by pairwise comparison method the criteria by significance are distributed as follows: durability is the most important criterion in the evaluation of alternatives; the price of a part of assembly process; construction workers' qualification level (category); mechanisation level of a part of assembling process; and complexity of assembling work are less important criteria.

Keywords: steel frame structure, technological solution, assembling work, network model, multiple criteria evaluation

1. Introduction

Modern construction industry is developing rapidly, thanks to new technologies. Steel frame structures are often used in the construction of public and industrial buildings. Buildings from insulated metal structures have the following advantages: excellent architectural look that meets the strictest requirements for modern buildings; simple and fast mounting enabling to

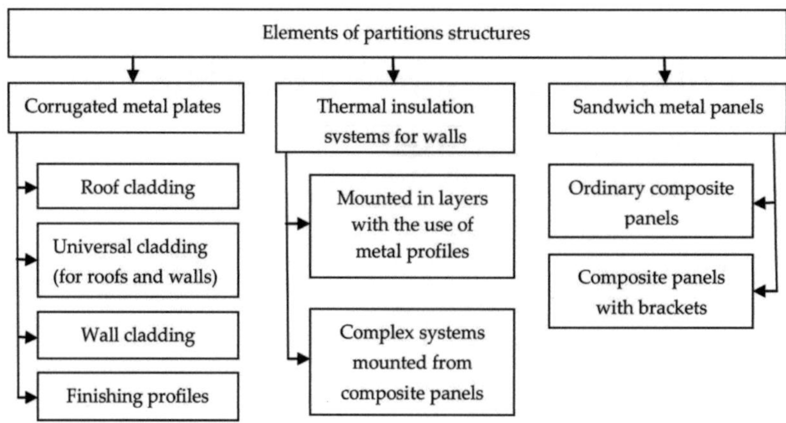

Figure 1. Elements of partition structures.

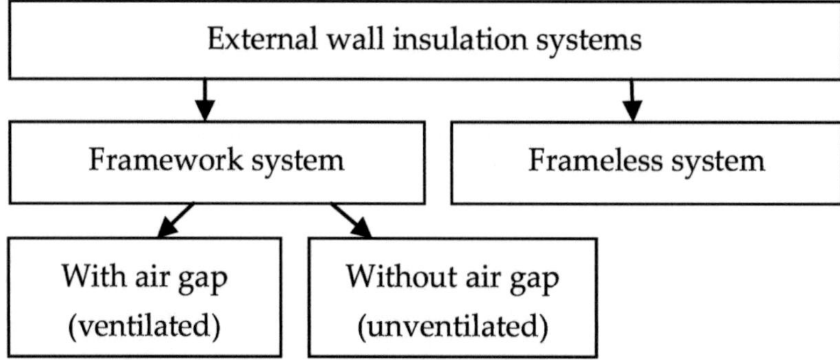

Figure 2. External wall insulation systems.

erect a big building in a very short time calculated in months or even in weeks; good thermal-technical characteristics as modern metal structures enable to avoid thermal bridging at connections, and the thickness of heat insulating layer is selected according to applicable standards; good operation characteristics as reliable elements ensure the required tightness of the building, and metal elements are protected from corrosion by more than one layer of coating.

Modern steel frames are made of the following components: bearing metal elements of the structural framework (partitions and load bearing elements); roof, roof-wall and wall cladding; partition insulation packages (thermal insulation, noise insulation, wind and vapour barriers). Elements of partitions structures and external wall insulation systems are presented in **Figures 1** and **2**.

Methods of insulating walls of existing industrial buildings:

- Thermal insulating layer is fixed directly to the façade that is afterward finished by reinforced (with mesh or fibre) plastering. It is a frameless insulation system.
- Wood or metal studs are fixed to the wall, thermal insulation is placed into the spaces between studs, and various finishing panels are fixed to the studwork. It is a framework system (ventilated).

- Finished elements made of joined thermal insulation, and finishing layers (composite panels) are fixed to the wall.

2. Designing the network model for alternative mounting solutions in steel frame building

2.1. Making combinations of complex processes

The main stages for designing network models for steel frame building are as follows:

- making combinations of complex processes used in steel frame building;
- finding possible alternatives of partial processes in steel frame building;
- finding technological links between the alternatives of partial processes in steel frame building;
- drawing networks for steel frame building technology.

In terms of system approach, steel frame building is a complex process made of various partial (work) processes that can be completed by different working methods, which are determined by work object characteristics, construction materials, work tools, equipment, number of workers and their qualification. Each of the above factors is described by certain technical and economic indicators [1–3].

The complex process of steel frame building can be divided into the following partial technological processes: F—steel frame mounting, B—mounting of bearing structures (beams and trusses); I—installing connections (**Figure 3**).

The complex process of wall and roof erection can be divided into the following partial technological processes: P—mounting of purlins, IL—inner layer mounting, S—sound insulation mounting, T—thermal insulation mounting, W—wind insulation mounting, and FL—finishing layer (**Figures 4** and **5**).

2.2. Alternatives of partial processes in steel frame building

The analysis of steel frame building reveals many technological systems of this complex process. Many alternative solutions can be found by changing work methods. Different work methods result from the change of building structures, work tolls and mechanisms used.

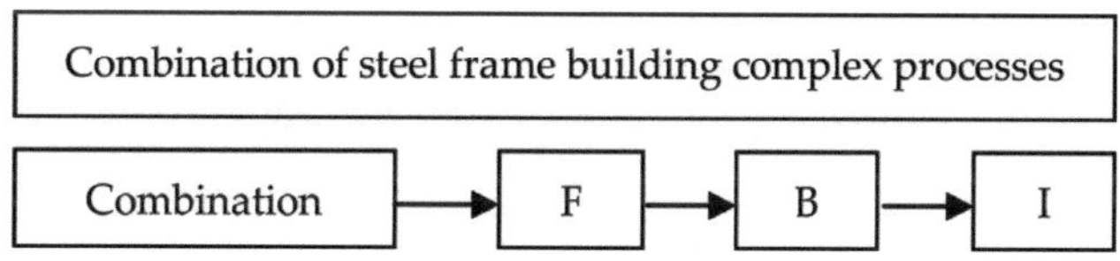

Figure 3. Diagram of steel frame building complex process combination.

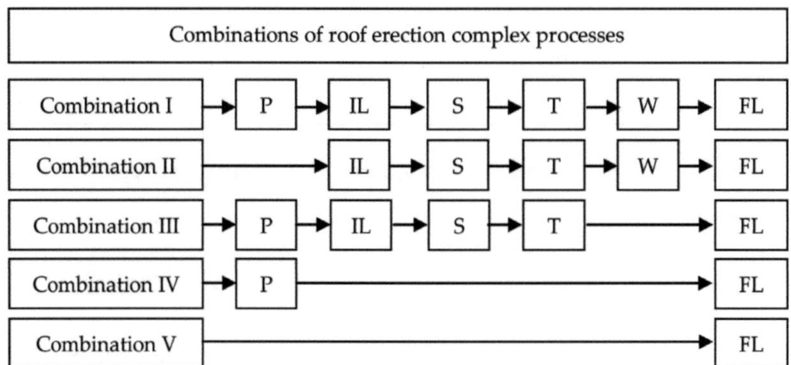

Figure 4. Diagram of wall erection complex process combination.

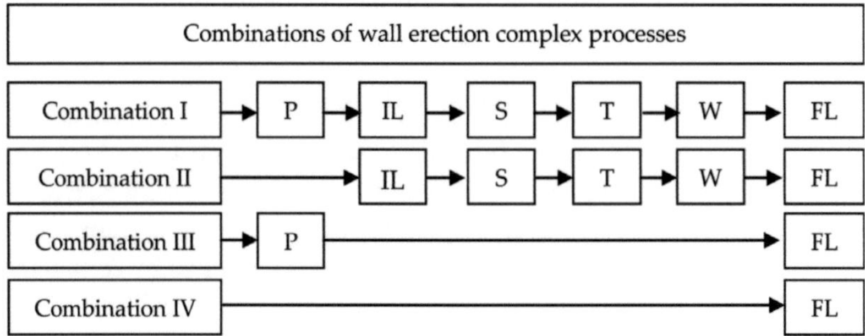

Figure 5. Diagram of roof erection complex process combination.

Available alternatives of work processes used for the building steel framework are presented in **Table 1**.

2.3. Technological links between partial processes in steel frame building and network model design

Network modelling of construction processes significantly improves operations management, work culture and efficiency, shortens the commissioning term, and reduces construction costs [4].

The method is beneficial only if the following conditions are met: well-organised data collection, transfer and processing; expedient system of decision-making and task delegation to operators supported by computerised network and specialists highly competent in network planning and control.

The graphical representation of the variety of works in a construction process with marked technological and organisational links is called a network model [5]. The network model with computed space, time, and technological parameters is called a network. Alternatives of separate (partial) technological processes and dependency relationship between them must be

Partial process alternative code	Title of partial process in steel frame building: partial process alternatives (short description)
	Mounting of columns:
F1	Manually, bolted connections
F2	Manually, welding
F3	Mechanically, crane lifting, bolted connections
F4	Mechanically, crane lifting, welding
F5	Mechanically, hoist lifting, bolted connections
F6	Mechanically, hoist lifting, welding
	Mounting of bearing structures:
B1	Mounting beams manually, bolted connections
B2	Mounting beams manually, welding
B3	Mounting beams, crane lifting, bolted connections
B4	Mounting beams, crane lifting, welding
B5	Mounting beams, hoist lifting, bolted connections
B6	Mounting beams, hoist lifting, welding
B7	Mounting trusses, hoist lifting, assembling on the ground
B8	Mounting trusses at designated height, hoist lifting
B9	Mounting trusses, assembling on the ground, crane lifting
B10	Mounting trusses at designated height, crane lifting
	Making connections:
I1	Making connections manually, welding
I2	Making connections manually, bolting
I3	Making connections, hoist lifting, bolting
I4	Making connections, hoist lifting, bolting
I5	Making connections, crane lifting, welding
I6	Making connections, crane lifting, bolting

Table 1. Alternatives of work processes used for the building a metal framework.

set in the design of network technological model. Technological links are made for three building alternatives: steel frame (**Figure 6**), roof (**Figure 7**), and wall (**Figure 8**).

Technological network model for the complex installation of steel frame is shown in **Figure 9**.

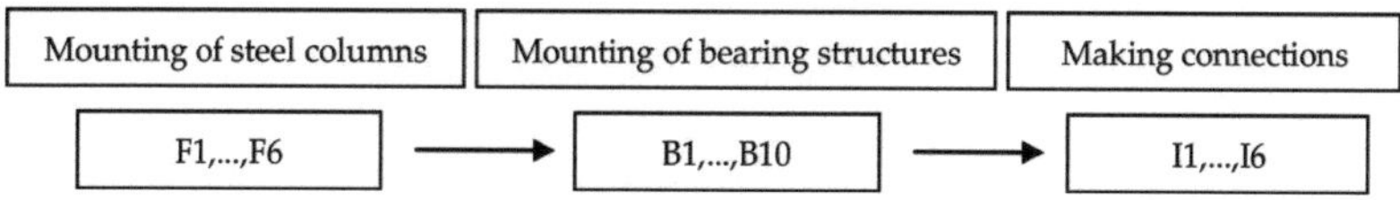

Figure 6. Technological links between partial processes in steel frame building.

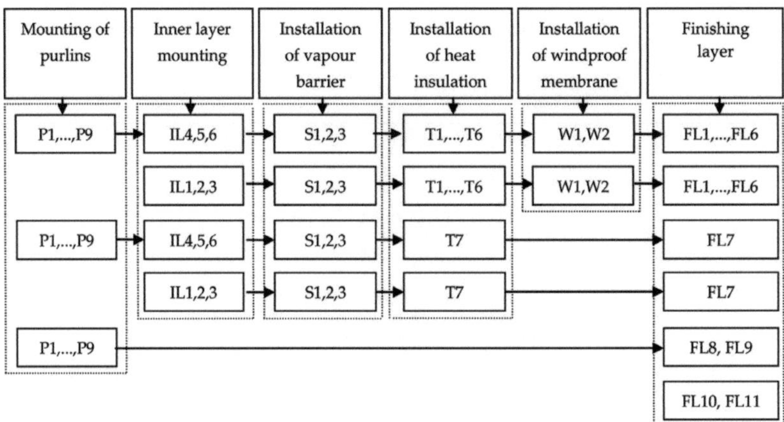

Figure 7. Technological links between partial processes in roof building.

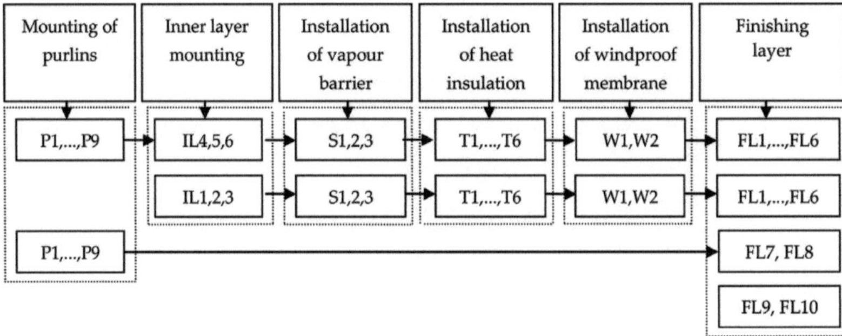

Figure 8. Technological links between partial processes in wall building.

3. Major principles of construction technological decision optimisation

Applicable technological and other project decision optimisation methods used in projection and construction processes could be divided into two major groups: applied mathematics and systemic-technical analysis methods. A great deal of construction organisation tasks could be solved by using mathematic statistics (correlation and regression analysis), theory of chances, mathematic programming, 'gambling' theory, multi-criteria optimisation and other methods.

The selection of optimisation method depends on the task character which is solved, the possessed source information and frequently it requires local interpretation. While solving practical construction optimisation tasks, most frequently only one (the most important) of several economic criteria is chosen (e.g. total construction price, object construction or separate pieces of construction per one solid metre (1 m^3) price, the revenue received, the greatest turnover, etc.). The significance of the criterion selected is very important. It shows that one of the criterions mentioned (e.g. revenue received), which is selected by the interested party, is

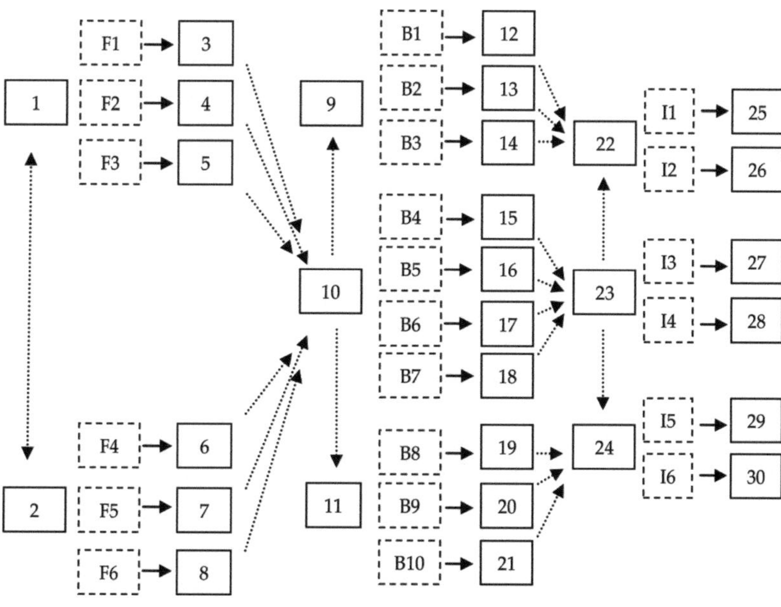

Figure 9. Technological network model for the complex installation of steel frame.

much more important than the other one, for example, total construction price. Therefore, to be concrete, it shows the significance of one criterion to the interested party in comparison with the rest, which are left as less significant or insignificant.

The significance of the criterion could be determined by employing statistical, expert opinion-based methods, even comparison, and entropy methods. These methods help to determine various theoretical, subjective, and complex values of great significance that are further used in the decision optimisation counting processes. By counting values, it is meant that the interested party could choose the criterion (they think) of the greatest significance.

Moreover, the client is frequently much more interested not in the price, but in other criteria such as construction duration, aesthetics, harmfulness to the health of the materials used, longevity, convenience to exploit, comfortability, and so on. So any construction technological decisions could be described and optimised according to the following system of criteria evaluation, where the criteria could be expressed by the indices of technological economy and quality characteristics. For this purpose, methods of multi-criteria decisions are used [1, 2].

It can be argued that each of the decision optimisation method mentioned has its own advantages and disadvantages. Moreover, each of them could be used to solve the tasks of specific constructions groups. They help to create various optimisation models of theoretical objects, technological or work processes like technological net models (alternative decisions, resources, dynamic, duration), mathematic models (shape of matrices, equation systems, various probability models), expert level systems, decision support systems and many others models.

Researches of modelling and optimisation were founded based on the applied mathematics method, economics, system theory, cybernetics, and in the sphere of counting technological science and its integration. While optimising technological construction processes, it is advisable to apply the theoretical principles of system methodology on the grounds that, nowadays, the technological projection methods which are used do not correspond to the requirements of effective decision-making. It could be noted that one of the major disadvantages are the decisions are accepted synonymically, without any preliminary examination and evaluation of the model of construction process technology and many the like multi-criteria evaluation [3]. While solving the any technological decision optimisation problem, it is necessary to perform three major steps of construction process systemic examination (**Figure 10**).

While modelling the construction composite process technological decisions, it is advisable to accept these main preconditions:

- all possible variations of complex process technologic decisions have to be constructed. Moreover, technological connections and partial variants of the processes (partial alternative decisions) have to be established;
- while forming the net model, consisting of partial process technological variants, it is necessary to take a precondition that only one of many other partial process technological variants will be implemented;
- every partial process has got its individual time duration, which is either technologically based or depends on the work expenditure.

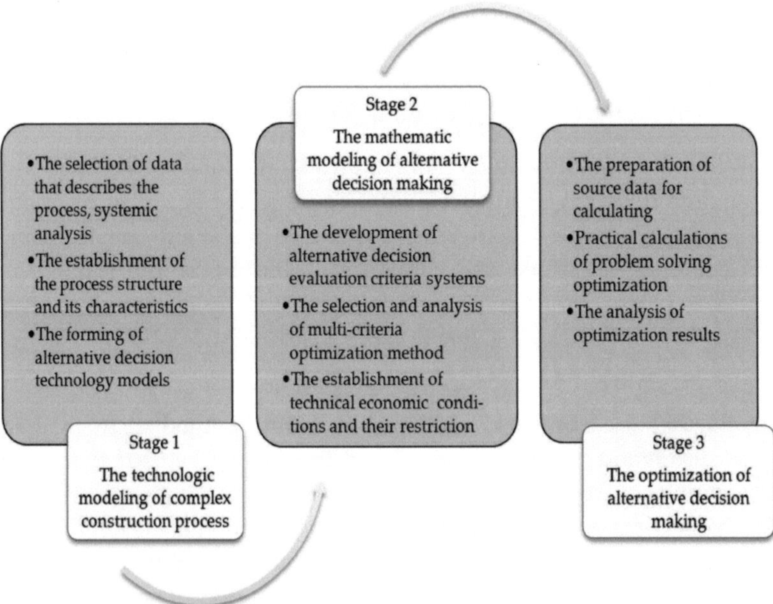

Figure 10. The steps of construction composite process systemic examination.

Alternative decisions	Evaluation criteria			
	K_1	K_2	…	K_n
a_1	x_{11}	x_{12}		x_{1n}
a_2	x_{21}	x_{22}		x_{2n}
…	…	…	…	…
a_m	x_{m1}	x_{m2}		x_{mn}

Table 2. The source data matrix *P*.

So while creating the mathematical model of alternative technological decision-making, it is advisable to define the set of the compared alternative decisions and their evaluation criteria. In that case, the source data matrix *P*, presented in **Table 2**, is prepared.

The source data matrix *P* most often consists of different units of measurement. For this reason, the matrix should be normalised, that is, it has to be transformed into the anti-dimensioned unit or sizes. Knowing the aims of the solution and applying the methods of normalisation, various normalised values of indices are obtained, which play the key role in other stages of solution in the field of multi-criteria optimisation.

The quantitative and qualitative characteristics are shown in **Table 3**.

Criteria	Unit of measurement	Definition
The price of a part of assembly process	(EUR/unit of measurement)	Price (in EUR) per conventional unit of measure of the analysed partial process alternative
The qualification level of construction workers	Category	Evaluation criterion indicating the workers' ability to do the work of the relevant complexity
Mechanisation level of a part of assembling process	%	
Durability	In year	Life cycle of steel frames
Complexity of assembling work	In points	Complexity of work evaluation criterion

Table 3. Alternatives of work processes used for the building a metal framework.

4. Major rudiments of applying the method of proximity to an ideal point used to evaluate the technology

The main essence of the multi-criteria evaluation method is the formation of generalised composed criterion. It is based on the comparison deviation of the criteria from so-called the ideal criteria, consisting of the best variant criteria being analysed. By applying the method and K_{bit} criteria, it is advisable to consider that each variant of the task problem solving utility function has the tendency to monotonously increase or monotonously decrease, that is, the larger value of any indices, the better it is or worse for less of the same index value. It depends on the fact whether the utility function increases or decreases. Indices have to be either cardinal or ordinal. If there are ordinal (qualitative) indices, they should be quantified. Besides, significance values should be determined, otherwise, they all are accepted as being equals. The application algorithm of the method of proximity to an ideal point, estimating the significance of the criteria, is presented in **Figure 11**.

The matrix P of alternative architectural decisions is created. There could also be criteria either grouped or ungrouped. The matrix normalisation is being done, according to the formula:

$$\overline{x_{ij}} = \frac{x_{ij}}{\sqrt{\sum_{i=1}^{m} x_{ij}^2}}, \tag{1}$$

where $i = \overline{1, m}; j = \overline{1, n}$.

If the significance of the subjective or theoretical ($\overline{q}$ or q_t) criteria is known, then the vector column multiplied by the normalised matrix corresponding column.

$$\text{Weighed matrix is obtained } \overline{P^*} = \left[\overline{P}\right] \cdot [q] \tag{2}$$

If there are no values of significance, then $\overline{P} = \overline{P^*}$ ($\overline{P}$ matrix is compared to the weighed matrix), that is, we take the precondition that entire alternative solution criteria are equally important. The ideal positive variant is being established:

$$a^+ = \left\{ \left[\left(\max_i f_{ij} / j \in I \right), \left(\min_j f_{ij} / j \in I' \atop j \right) \right] / i = \overline{1, m} \right\} = \{ f_1^+, f_2^+, \dots, f_n^+ \} \tag{3}$$

where I is the indices of ratio (maximising), which possesses the highest values.

The ideal negative variant is being established:

$$a^- = \left\{ \left[\left(\min_i f_{ij} / j \in I \right), \left(\max_j f_{ij} / j \in I' \atop j \right) \right] / i = \overline{1, m} \right\} = \{ f_1^-, f_2^-, \dots, f_n^- \} \tag{4}$$

The difference (distance) between real and ideal positive variant is being found:

$$L_i^+ = \sqrt{\sum_{j=1}^{n} \left(f_{ij} - f_j^+ \right)^2} \tag{5}$$

where a_i is the real variant, a^+ is the ideal positive variant, and L_i^+ is the positive distance.

The difference between real and ideal negative variant is being found:

$$L_i^- = \sqrt{\sum_{j=1}^{n} \left(f_{ij} - f_j^-\right)^2} \tag{6}$$

$K_{bit,i}$ calculation of values (each alternative value is found):

$$K_{bit,i} = \frac{L_i^-}{L_i^+ + L_i^-}, \text{when } \forall_i; \ \ i = \overline{1, m}. \tag{7}$$

$0 \leq K_{bit} \leq 1$, besides,

$$K_{bit,i} = \begin{Bmatrix} 1, jei & a_i = a^+ \\ 0, jei & a_i = a^- \end{Bmatrix} \tag{8}$$

The best (the most rational) architectural solution will become the one, which K_{bit} value will be max ($K_{bit,i}$ = max). Using the values generates the priority sequence utility degree establishment. The value of the variant being tested is compared to the value of the ideal variant.

$$N_i = \frac{K_{bit,i}}{K_{bit,\max}} \cdot 100\% \tag{9}$$

The method of proximity to an ideal point can be applied to find the most effective engineering solution alternative.

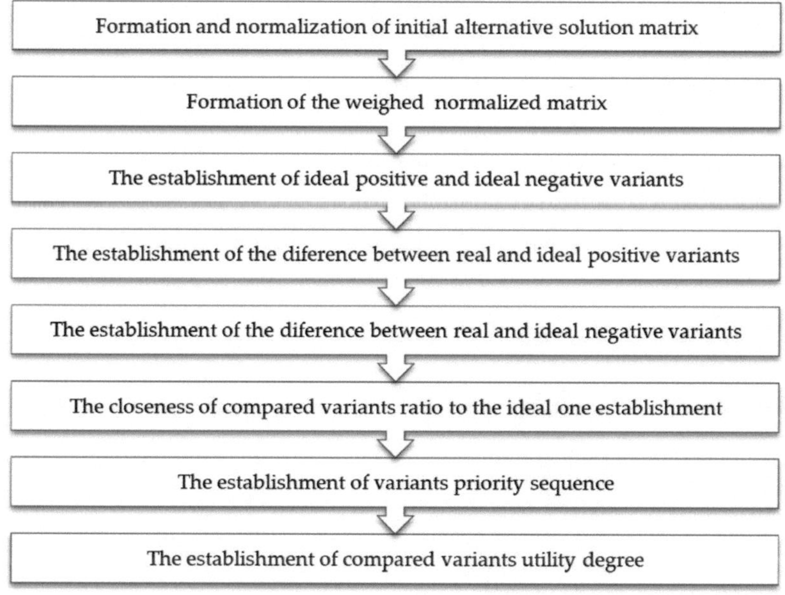

Figure 11. Application algorithm of the method of proximity to an ideal point.

5. Multi-criteria evaluation of alternative solutions in steel frame building

The rational alternative for steel frame building is found from the diagram presented in **Figure 12**.

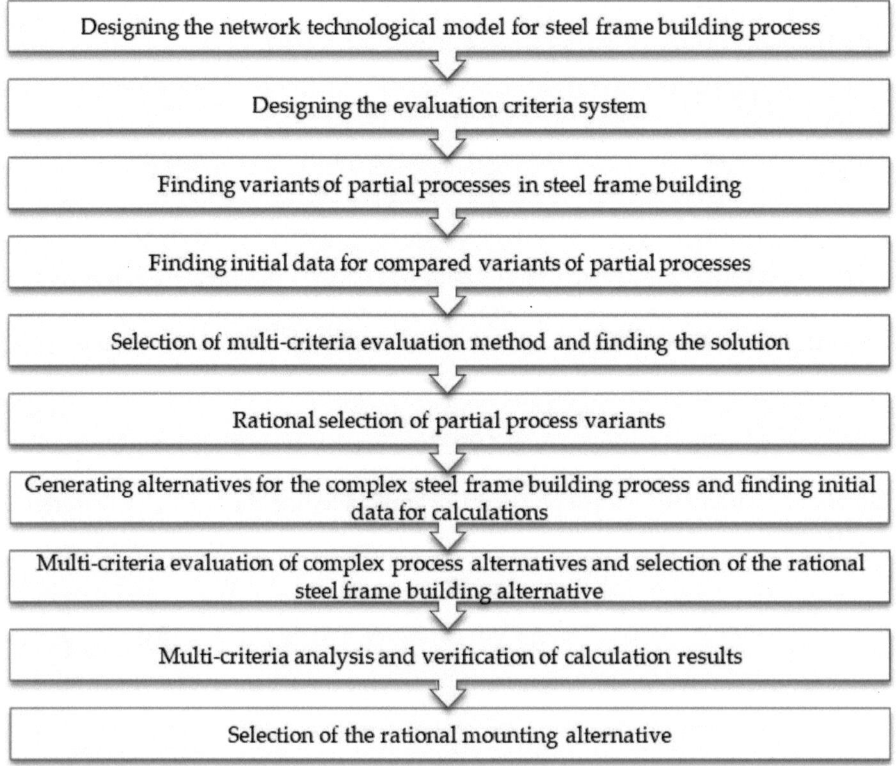

Figure 12. Algorithm for the selection of the rational alternative for the complex process.

6. Conclusions

Taking into consideration the factors that affect the rationality of steel frame building process solutions, it is feasible to do the technological modelling of multi-criteria evaluation of alternative buildings.

A criteria system for the evaluation of alternative partial processes must be designed in order to find the rational steel frame building alternative. Criteria values and importance may be subsequently adjusted taking priorities and the current situation into account.

In practice, it is possible to find the most rational technological alternatives for metal framework, roof and wall structures separately with the help of network technological model and multi-criteria analysis of steel frame building solutions.

http://dx.doi.org/10.5772/intechopen.79858

In multiple criteria evaluation of technological solutions for assembling buildings from steel frame structures by pairwise comparison method, the criteria by significance are distributed as follows: durability is the most important criterion in the evaluation of alternatives; the price (EUR/unit of measurement) of a part of assembly process; construction workers' qualification level (category); mechanisation level of a part of assembling process (%); and complexity of assembling work (in points) are less important criteria.

Conflict of interest

The author declares no conflict of interest.

Author details

Ruta Miniotaite

Address all correspondence to: rutaminiot@gmail.com

Kaunas University of Technology, Kaunas, Lithuania

References

[1] Zavadskas EK, Kaklauskas A, Banaitiene N. Multiple Analysis of the Life of the Building Process. Vilnius: Technica Press; 2001

[2] Zavadskas EK, Kaklauskas A. Systematic and Technical Evaluation of the Buildings. Vilnius: Technica Press; 1996

[3] Zavadskas EK, Simanauskas L, Kaklauskas A. Dacision Support System in Construction. Vilnius: Technica Press; 1999

[4] Juodis A. Modeling and Optimization of Construction Processes. Kaunas: Technology Press; 2005

[5] Ginevicius R, Podvezko V. A feasibility study of multicriteria methods' application to quantitative evaluation of social phenomena. Business. 2008;**9**(2):84-87